THE BADGER

Proceedings of a seminar held on 6-7 March 1991

edited by

Thomas J. Hayden

Department of Zoology, University College, Dublin

Royal Irish Academy
Dublin 1993

ROYAL IRISH ACADEMY
19 DAWSON STREET
DUBLIN 2

ISBN 1 874045 14 3

British Library Cataloguing-in-Publication Data.
A catalogue record for this book is available from the British Library.

Also published by the Royal Irish Academy

Renewable resource development in Ireland
edited by A.C. Cassells

Biological indicators of pollution
edited by D.H.S. Richardson

Taxonomy: putting plants and animals in their place
edited by Christopher Moriarty

Ecology and conservation of Irish peatlands
edited by Gerry Doyle

Irish rivers: biology and management
edited by Martin Steer

Plant health and 1992
edited by J.A. Kavanagh and P. Brennan

CONTENTS

This volume is dedicated to the memory
of
Asia Ibrahim Il-Fituri

LIST OF CONTRIBUTORS

BENHAM, P. F. J.	Primrose Farm, Felindre, Brecon, Powys, UK.
BROWN, J. A.	Department of Zoology, University of Bristol, UK.
CHEESEMAN, C. L.	Central Science Laboratory, Ministry of Agriculture, Fisheries and Food, Guildford, Surrey, UK.
DOLAN, L.	TB Investigation Unit, University College, Dublin, Ireland.
DUKE, E. J.	Department of Zoology, University College, Dublin, Ireland.
ELGADI, A. S.	Department of Zoology, University College, Dublin, Ireland.
EVES, J.	District Veterinary Office, Department of Agriculture and Food, Tullamore, Co. Offaly, Ireland.
FAGAN, J.	Regional Veterinary Laboratory, Coosan, Athlone, Co. Westmeath, Ireland.
FEORE, S.	School of Biology and Biochemistry, Queen's University, Belfast, Northern Ireland.
GOODGER, J.	Central Veterinary Laboratory, Weybridge, Surrey, UK.
FLANAGAN, P.	District Veterinary Office, Tullamore, Co. Offaly, Ireland.
HARRIS, S.	Department of Zoology, University of Bristol, UK.
HAYDEN, T. J.	Department of Zoology, University College, Dublin, Ireland.
IL-FITURI, A. I.	Department of Zoology, University College, Dublin, Ireland.
KIRK, S.	Minister of State at the Department of Agriculture and Food, Agriculture House, Dublin 2, Ireland.
KRUUK, H.	Institute of Terrestrial Ecology, Banchory, Kincardineshire, UK.
LYNCH, J. M.	Department of Zoology, University College, Dublin, Ireland.
MCCARTHY, J.	District Veterinary Office, Department of Agriculture and Food, Cork, Ireland.
MAHMOOD, K. H.	Department of Animal Production, University of Basrah, Iraq.
MALLINSON, P. J.	Central Science Laboratory, Ministry of Agriculture, Fisheries and Food, Guildford, Surrey, UK.
MONTGOMERY, W. I.	School of Biology and Biochemistry, Queen's University, Belfast, Northern Ireland.
MULCAHY, M. F.	Department of Zoology, University College, Cork, Ireland.
NOLAN, A.	Central Veterinary Laboratory, Weybridge, Surrey, UK.

O'CORRY-CROWE, G.	Department of Zoology, University College, Dublin, Ireland.
POWER, S. B.	Regional Veterinary Laboratory, Department of Agriculture and Food, Cork, Ireland.
ROPER, T. J.	School of Biology, University of Sussex, Brighton, UK.
RYAN, J.	Central Veterinary Laboratory, Weybridge, Surrey, UK.
SLEEMAN, D. P.	Department of Zoology, University College, Cork, Ireland.
SMAL, C. M.	The Wildlife Service, Office of Public Works, Bray, Co. Wicklow, Ireland.
STAINSBY, K.	Department of Microbiology, University of Surrey, Guildford, UK.
STANFORD, J.	Department of Medical Microbiology, University College, London, UK.
WARNER, P.	The Wildlife Service, Office of Public Works, Bray, Co. Wicklow, Ireland.
WHELAN, R.	Department of Zoology, University College, Dublin, Ireland.
WHELTON, H. J.	Department of Zoology, University College, Cork, Ireland.
WILESMITH, J. W.	Central Veterinary Laboratory, Weybridge, Surrey, UK.

LIST OF PARTICIPANTS

Mr R. Allen,	District Veterinary Office, Sligo
Mr L. Armstrong,	District Veterinary Office, Sligo
Dr J. M. Barry,	Sidney Place, Cork
Dr P. F. J. Benham,	University of Reading
Mr J. Bradley,	Regional Veterinary Laboratory, Limerick
Mr C. Brennan,	Irish Wildlife Federation, Dublin
Dr P. J. Brennan,	Faculty of Veterinary Medicine, University College, Dublin
Ms C. Brookman,	Badgerwatch Ireland, Dublin
Mr J. A. Brown,	Department of Zoology, University of Bristol, UK
Mr J. Cahill,	Department of Zoology, University College, Dublin
Mr E. Callaghan,	Department of Zoology, University College, Dublin
Mr M. Canny,	Wildlife Service, Bray, Co. Wicklow
Dr C. L. Cheeseman,	UK Ministry of Agriculture Fisheries and Food
Mr W. Clarke,	Department of Zoology, University College, Dublin
Mr L. A. Clements,	Department of Agriculture for Northern Ireland
Mr L. Clements,	Regional Veterinary Laboratory, Sligo
Professor D. Collins,	Faculty of Veterinary Medicine, University College, Dublin
Mr E. Costello,	Veterinary Research Laboratory, Dublin
Mr P. J. Crowe,	Regional Veterinary Laboratory, Kilkenny
Mr H. Cummins,	District Veterinary Office, Cork
Mr G. O. Denny,	Department of Agriculture for Northern Ireland
Mr P. J. A. Doherty,	Department of Agriculture and Food, Sligo
Mr T. Doherty,	Badgerwatch Ireland, Dublin
Dr L. Dolan,	Faculty of Veterinary Medicine, University College, Dublin
Mr E. Doran,	Wildlife Service, Bray, Co. Wicklow
Dr P. Dowding,	Trinity College, Dublin
Dr L. Downey,	ERAD, Department of Agriculture and Food, Dublin
Mr M. Duggan,	District Veterinary Office, Nenagh
Professor E. J. Duke,	Department of Zoology, University College, Dublin
Mr L. Egan,	Irish Farmers Association, Dublin
Mr A. Elgadi,	Department of Zoology, University College, Dublin
Mr J. Eves,	District Veterinary Office, Tullamore
Mr J. Fagan,	Regional Veterinary Laboratory, Athlone
Mrs M. Feore,	Department of Agriculture for Northern Ireland
Ms S. Feore,	School of Biology and Biochemistry, Queen's University, Belfast
Mr P. Flanagan,	District Veterinary Office, Tullamore
Ms E. Giddy,	Wildlife Service, Co. Carlow
Mr E. Grennan,	Wildlife Service, Bray, Co. Wicklow
Mr J. Griffin,	Faculty of Veterinary Medicine, University College, Dublin
Mr B. Halvey,	Irish Farmers Association, Dublin
Dr R. Hammond,	Faculty of Veterinary Medicine, University College, Dublin
Ms A. Hanlon,	Department of Zoology, University College, Dublin
Dr R. Harrington,	Wildlife Service, Bray, Co. Wicklow
Dr S. Harris,	Department of Zoology, University of Bristol, UK
Dr T. Hayden,	Department of Zoology, University College, Dublin
Mr M. L. Hayes,	Department of Agriculture and Food, Nenagh
Mr B. Hennessey,	District Veterinary Office, Carlow

Ms A.I Il-Fituri,	Department of Zoology, University College, Dublin
Professor J. A. Kavanagh,	University College, Dublin
Mr J. Kehoe,	Irish Wildlife Federation
Mr A. Kelly,	Faculty of Veterinary Medicine, University College, Dublin
Mr P. Kelly,	Department of Zoology, University College, Dublin
Dr H. Kruuk,	Institute of Terrestrial Ecology, Banchory, Scotland
Ms F. Lang,	Department of Zoology, University College, Dublin
Dr J. Larner,	Killarney National Park
Mr J. Lynch,	Department of Zoology, University College, Dublin
Mr P. McAleer,	District Veterinary Office, Galway
Mr P. McArdle,	Department of Agriculture and Food, Mullingar
Mr J. McCarthy,	District Veterinary Office, Cork
Ms K. McGill,	District Veterinary Office, Roscommon
Mr J. McManus,	Battersea, London, UK
Mr D. Maguire,	Irish Farmers Journal, Dublin
Mr P. J. Mallinson,	UK Ministry of Agriculture
Mr F. Maughan,	District Veterinary Office, Belmullet, Co. Mayo
Ms R. Meleady,	Irish Wildlife Federation, Dublin
Mr P. Miller,	Glenveagh National Park
Mr J. Moynagh,	Department of Agriculture and Food, Dublin
Ms P. Murphy,	Department of Zoology, University College, Dublin
Dr D. Murray,	Department of Zoology, University College, Dublin
Dr S. Neill,	Veterinary Research Laboratory, Dublin
Dr S. Neill,	Queen's University, Belfast
Dr A. Nolan,	Central Veterinary Laboratory, UK
Mr D. Nolan,	Department of Zoology, University College, Dublin
Mr T. J. O'Brien,	District Veterinary Office, Roscommon
Mr A. J. O'Connell,	Department of Agriculture and Food, Limerick
Mr D. O'Connell,	Department of Agriculture and Food, Carlow
Ms J. O'Connell,	Department of Zoology, University College, Dublin
Mr G. O'Corry-Crowe,	Department of Zoology, University College, Dublin
Ms A. O'Donoghue,	Department of Zoology, University College, Dublin
Mr J. O'Donoghue,	Wildlife Service, Bray, Co. Wicklow
Mr T. P. O'Hanlon,	District Veterinary Office, Carlow
Mr D. O'Hare,	Irish Wildlife Federation
Mr C. O'Keeffe,	Glenveagh National Park
Mr J. O'Keeffe,	Department of Agriculture and Food, Dublin
Ms N. O'Keeffe,	Wildlife Service, Dublin
Mr S. O'Neill,	Irish Farmers Association, Dublin
Dr L. O'Reilly,	Veterinary Laboratory, Dublin
Mr P. F. O'Reilly,	Veterinary Research Laboratory, Dublin
Mr. B. Ó Suilleabháin,	District Veterinary Office, Sligo
Mr P. O'Sullivan,	Wildlife Service, Ennis, Co. Clare
Ms E. D. Overend,	Chippenham, Wiltshire, UK
Mr T. Partridge,	Department of Agriculture and Food, Co. Waterford
Mr L. Power,	Irish Wildlife Federation, Dublin
Mr S. Power,	Regional Veterinary Laboratory, Cork
Mr P. Prendeville,	Faculty of Veterinary Medicine, University College, Dublin
Ms T. Reilly,	University College, Dublin
Dr T. J. Roper,	University of Sussex, UK
Ms M. Ryan,	University College, Dublin

Mr C. F. Saich,	Wildlife Service, Bray, Co. Wicklow
Ms M. J. P. Scannell,	Ballsbridge, Dublin
Mr P. Sheridan,	Veterinary Research Laboratory, Dublin
Dr D. P. Sleeman,	Department of Zoology, University College, Cork
Dr C. Smal,	Wildlife Service, Bray, Co. Wicklow
Mr P. J. Smiddy,	Wildlife Service, Cork
Professor J. Stanford,	Middlesex Hospital, UK
Mr P. Timoney,	District Veterinary Office, Cavan
Ms A. Tinney,	Badgerwatch Ireland, Dublin
Mr P. Tully,	District Veterinary Office, Sligo
Mr M. A. Twomey,	Derryharrow, Longford
Mr M. Viney,	The Irish Times
Mr P. Warner,	Wildlife Service, Bray, Co. Wicklow
Dr J. Whelan,	Department of Environmental Resource Management, University College, Dublin
Ms R. Whelan,	Department of Zoology, University College, Dublin
Ms H. J. Whelton,	Department of Zoology, University College, Cork
Mr J. Wilson,	Wildlife Service, Bray, Co. Wicklow
Ms J. Woodlock,	Glenveagh National Park

ACKNOWLEDGEMENTS

The editor, on behalf of the National Committee for Biology of the Royal Irish Academy, is pleased to acknowledge the generous support of the Wildlife Service, the Animal Diseases Eradication Board (ERAD) and University College, Dublin.

The success of the seminar was due to the excellent efforts of Ms Sara Whelan of the Royal Irish Academy and Dr Declan Murray, Department of Zoology, University College, Dublin.

The chairpersons and panel of referees are also warmly thanked for their efforts during and after the seminar.

Special thanks are due to Rosemary Whelan, Gregory O'Corry-Crowe and John Lynch, who typed much of the text and prepared the camera-ready copy for this volume.

In: Hayden, T.J. (ed.) 1993 *The badger*, 1. Dublin. Royal Irish Academy.

INTRODUCTION

T.J. Hayden

Mammal Research Group, Department of Zoology, University College, Dublin

In recent years the badger has become an animal of major scientific and public interest in Ireland. There has been a marked increase both in interest in the possible role of the badger in the aetiology of bovine tuberculosis and in the activity of groups interested in the conservation of the badger. The purpose of this seminar is to bring together current information on (1) the biology of the badger in Ireland, (2) the ongoing investigations of the aetiology of tuberculosis in bovines in so far as they concern the badger, and (3) recent research overseas which is relevant to an understanding of the biology of the badger and the control of tuberculosis.

Although the badger is Ireland's largest extant carnivore, less than five scientific publications on the biology of this animal in Ireland have been published in the last two decades.

This seminar addresses this deficiency in that it brings together all groups currently carrying out research on the badger in this country. It is intended, by placing available information in the public domain, to identify where further information is required and to encourage discussion and collaboration between interested partners so that further relevant research is stimulated.

Some of the reports concern work in progress. These are included because, as they necessarily represent longer-term studies, it is appropriate that the background and hypotheses under investigation be clearly stated and, more importantly, how the results are to be analysed and the time-scale involved.

All the papers were read by at least two reviewers and were modified, some substantially, as a result. The final decision rested with the editor, in the light of the stated objectives of the seminar.

The organising committee and the editor wish sincerely to thank all those who have contributed to the organisation of the seminar and the production of this volume.

In: Hayden, T.J. (ed.) 1993 *The badger*, 2-3. Dublin. Royal Irish Academy.

OPENING ADDRESS

S. Kirk, T.D.

Minister of State at the Department of Agriculture and Food

The badger is a mammal which has of late attracted quite an amount of attention in this country and also in the UK.

The Wildlife Service of the Office of Public Works is the national authority with responsibility for the protection and conservation of wildlife in this country. Because of the relatively low number of native species in Ireland, the elements of protection and conservation are particularly important. There are some seven indigenous mammal species in all, namely, the red deer, the pine marten, the badger, the otter, the fox, the hare and the pygmy shrew. Each of these is protected by the provisions of the Wildlife Act 1976, under which it is an offence to deliberately interfere with them except where a licence is issued for a purpose specified under the Act. These animals are also protected species under the Bern Convention on the Conservation of European Wildlife and Natural Habitats, to which Ireland is a contracting party. The provisions of the Convention are expressed in Ireland within the scope of the Wildlife Act 1976.

Ireland, therefore, has both national and international responsibilities relating to the custodianship of these species. This also ties in with the preservation of habitats and the concept, now gaining increasing support, that both agriculture and industry must operate in an environmentally friendly fashion. Equally, it is recognised that a balanced approach is necessary. Our cattle and dairy sectors contribute a huge amount to export earnings. With the arrangements for the completion of the Single Market in the offing, together with the anticipated new stringent marketing restrictions pertaining to produce originating from herds with bovine TB, the Eradication of Animal Diseases Board (ERAD) is intensifying the effort to eliminate the remaining pockets of the disease. The government, recognising our major dependence on agricultural exports, continues to commit very significant financial and staff resources to disease eradication. Consumers are becoming more and more concerned with the health status of the food sources. This is particularly true in the European Market where we must seek an increasing market share, and also to a varying degree in other countries to which we export.

It is increasingly accepted that the badger is associated with TB infection in cattle; the question requiring clarification is the degree to which the badger is a spreading agent. We should, however, emphasise that in no sense should the badger be regarded as the sole cause of the spread of infection. The factual position is that, over the 40-odd years of the operation of the Bovine Tuberculosis Scheme, a number of sources of infection have been identified and monitored.

Research is obviously one of the essential elements in achieving a successful conclusion for the eradication strategy. Part of the current research programme is focused on defining the optimum badger population for Ireland. The National Badger Survey initiated by ERAD in conjunction with the Wildlife Service should provide an accurate assessment of existing populations. The Survey will considerably increase our knowledge of the species in terms of habitats and density patterns. The Road Casualty Survey which has recently been initiated will help clarify the present level of infection in badgers. In the long term, the aspiration is that research will discover a means of effective vaccination to ensure the health status of the badger species. I believe that a proper balance can be struck between preserving a large and healthy population and achieving the objectives of the disease eradication programme.

The badger has long been a resident of our countryside. References to '*an broc*' are evident in early Irish literature and in place-names such as Clonbrock in Co. Galway, Meenabrock in Co. Donegal, and Brockagh Mountain in Co. Wicklow. As I have indicated, there is an increasing awareness of the value and importance of wildlife and the environment in general. The various conservation organisations have an important role to play in highlighting issues of interest to the public. These organisations are a forum for the exchange of information, identification of problems, and for bringing matters to the attention of the government.

In conclusion, I would like to emphasise that my Department, which includes ERAD, will continue, in conjunction with the Wildlife Service, to seek ways and means of ensuring a healthy badger population.

Go raibh míle maith agaibh. Rath Dé ar bhur saothair.

In: Hayden, T.J. (ed.) 1993 *The badger,* 4. Dublin. Royal Irish Academy.

THE ORGANISATION OF BADGER POPULATIONS

H. Kruuk

Institute of Terrestrial Ecology, Banchory, Scotland

SUMMARY

From an overview of the highly variable organisation of the European badger, its evolution, and environmental factors which may affect it, it is argued that an understanding of the population organisation is important for the purpose of epidemiology and species management.

Most species of the mustelid family are solitary, with females defending small territories against other females, and males in larger territories, defended against males but overlapping with females. The organisation of some species has evolved into greater complexity, and the European badger is probably the most social. This gregariousness is related to the species feeding ecology.

Badger populations consist of groups of 'clans', each containing several males and females (up to 20 or more), defending strict boundaries around a clan territory. Numbers of adults in each clan can be determined using a harmless radioisotope, Zn^{65}, and it is shown that the size of each clan is related to food availability. Behavioural relations between badgers indicate that clan territories are scent-marked and defended against other badgers. The size of these territories was determined with radio-tracking and the recovery of coloured bait-markers from the faeces on territorial boundaries; it varied from 20 to 1500 hectares. The variation in territory size was not related to numbers of badgers in the clan; however, it appeared to be correlated with measures of dispersion of the food supply. The badger diet and foraging behaviour varies considerably, but the most important prey in north-west Europe is earthworms, and characteristics of the earthworm populations can explain many of the variables of badger organisation and numbers.

In: Hayden, T.J. (ed.) 1993 *The badger*, 5-8. Dublin. Royal Irish Academy.

BADGERS AND THE LAW

P. Warner

Wildlife Service, Bray, Co. Wicklow

THE WILDLIFE ACT 1976

The relevant piece of national wildlife legislation covering badgers in Ireland is the Wildlife Act 1976. This is a comprehensive law and awards differing degrees of protection to all the native mammals, birds, reptiles and amphibians (Forest and Wildlife Service 1980).

The mammals are divided into three categories.

1. 'Pest' species are given no status or protection under the Act. These are the brown rat (*Rattus norvegicus*), the black rat (*Rattus rattus*), the house mouse (*Mus domesticus*), the grey squirrel (*Sciurus carolinensis*), the mink (*Mustela vison*), the fox (*Vulpes vulpes*), the rabbit (*Oryctolagus cuniculus*), the bank vole (*Clethrionomys glareolus*) and the woodmouse (*Apodemus sylvaticus*).
2. The protected species may not be hunted, killed, nor their breeding places disturbed. These are whales and dolphins (all members of the order Cetacea), the grey seal (*Halichoerus grypus*), the common seal (*Phoca vitulina*), other members of the order Pinnipedia, bats (currently 7 species), the hedgehog (*Erinaceus europaeus*), the stoat (*Mustela erminea*), the otter (*Lutra lutra*), the red squirrel (*Sciurus vulgaris*), the pine marten (*Martes martes*), the pygmy shrew (*Sorex minutus*) and of course the badger (*Meles meles*).
3. The game species are protected but may be hunted in accordance with the sections specifying open seasons, type of permitted firearms, etc. These species are the Irish hare (*Lepus timidus*), the brown hare (*Lepus europaeus*), the fallow deer (*Dama dama*), the red deer (*Cervus elaphus*), the sika deer (*Cervus nippon*) and red x sika hybrids.

The legislators who drew up the list in 1976 considered the badger to be a species which was not a significant pest to agriculture, forestry or game management. Although the list has been amended since 1976, it has never been proposed to change the legal status of the badger.

DEROGATIONS

The 1976 Act recognises that protected species may from time to time damage man's interests. The classic example would be a stoat killing poultry.

A procedure exists under Section 42 of the Wildlife Act to license the capture, removal or killing of individual animals which are clearly doing damage in these cases. The Act also recognises that sometimes, as in the example above, there is no time to apply for a licence, and a 'good defence of haste' exists whereby the farmer can deal with the stoat on the spot and defend his actions in court with a reasonable probability that the Justice will accept his defence. The 'good defence of haste' is not permitted for some rare species (pine marten, red deer). The badger is not included in this list and so was not considered by the legislators to be exceptionally rare or endangered.

LICENCES TO REMOVE BADGERS

In the period since the Act was passed (1976-1990), only a few Section 42 licences to evict or kill badgers were granted, mainly for cases where the badger's digging activities were causing damage. In the 1980s the Department of Agriculture began to make applications for licences linked with the investigation of disease, and the number of applications has substantially increased since then (Table 1). Initially these were granted under Section 42, but subsequently this was changed to Section 23 which permits killing for 'educational or scientific purposes' but not for the control of damage. This change represented the view of the Wildlife Service that the link between badgers and TB (and hence the damage) was not proven but was still subject to investigation. This remains the position of the Wildlife Service. In 1990, 288 applications, some multiple, were received from ERAD. Currently licences are still granted under Section 23, each covering an area within 2km of a named farm. The licences do not provide legal powers to enter on land, which must be by permission of the landowner or by powers granted under Department of Agriculture legislation.

Table 1. Licensed capture of badgers under the Wildlife Act 1976.

Year	Applications per year, n	Relevant section of Act*	Badgers caught	No. with TB lesions
1977-1986	2-3	42	n.a.**	n.a.
1987	75	42	n.a.	n.a.
1988	172	42	1307	199
1989	270	23	519	76
1990	289	23	1223	148
1991	633	23	2780	280

*Section 42 relates to damage caused by badgers; Section 23 relates to capture for a scientific purpose.
** Statistics not available.

Licences under Section 23 are given on certification from the district veterinary officer, countersigned by his superior officer(s), that they are satisfied that there is valid reason to suspect that an outbreak of TB is attributable to badgers rather than to poor fencing or some other cause. The Wildlife Service reserves the right to make field inspections and to refuse licences. Licences are valid for one year.

INTERNATIONAL TREATY LAW

The badger is also covered in international treaty law in the sense that Ireland has ratified the Convention on the Conservation of European Wildlife and Natural Habitats (the Berne Convention). The aims of the Convention are to 'conserve wild flora and fauna and their natural habitats' and to 'give particular emphasis to endangered and vulnerable species'; in other words, the Convention does not concern itself particularly with the common or pest species. It has two lists, Appendix II (strictly protected animals) and Appendix III (protected animals). The species listed in Appendix III are afforded general protection by regulation of exploitation 'in order to keep the populations out of danger' (Article 7), and protection from indiscriminate methods of capture and killing (Article 8), in particular from stated means including snares, poison and gassing (Appendix IV). There is, however, a provision (Article 9) for signatory states to derogate from the provisions of all the above Articles in the interests of, amongst other things, the protection of livestock from 'serious damage'.

The badger appears in Appendix III, as do the seal species, pine marten and many cetaceans. This means that the consensus opinion over all the European range of the species (the EC countries, Switzerland, Austria, southern Scandinavia) is that the badger is basically harmless and worthy of protection. The badger was not included in Appendix II because it was not believed to be exceptionally rare or endangered and because it is hunted in some European states. Essentially, such species (Appendix III) are not perceived as requiring specific or active protection. Populations and distribution were at least adequate, i.e. they had a 'good' conservation status. The general measures applying to them in the Act and Convention are designed to protect their breeding places and, by a system of licences covering various methods of hunting and killing, to prevent over-exploitation.

OTHER NATIONAL LEGISLATION

The Diseases of Animals Act 1966 empowers the agents of the minister for agriculture to have slaughtered (Section 17) any 'animal' (animal is defined and includes all mustelids and hence badgers) found on land where bovine TB is present. Section 20 gives the minister powers to take possession of animals suspected of being disease-carriers. Section 25 empowers the minister's agents to 'take samples and apply tests', to enter on land and to command help from the landowner. These powers have not been used to date in the investigation of badgers and bovine TB.

Cruelty to animals is covered by the Protection of Animals Act 1911 as amended by the Protection of Animals (Amendment) Act 1965. These Acts, together with Section 34 of the Wildlife Act 1976, regulate the methods of trapping and of subsequently killing badgers. Currently a 'stopped snare' (that is, a snare that will not close beyond a certain diameter and so holds rather than strangles), with a swivel to prevent kinking, is the recommended method. Killing is by shooting or by lethal injection.

There has been much debate about the most humane and efficient methods of trapping badgers, but this is not within the scope of this paper.

REFERENCES

FOREST AND WILDLIFE SERVICE 1980 *Wildlife and the Law: A guide to the Wildlife Act, 1976* (2nd edn). Dublin. Stationery Office.

In: Hayden, T.J. (ed.) 1993 *The badger*, 9-22. Dublin. Royal Irish Academy.*

THE NATIONAL BADGER SURVEY: PRELIMINARY RESULTS FOR THE IRISH REPUBLIC

C. M. Smal

Wildlife Service, Office of Public Works, Bray, Co. Wicklow

ABSTRACT

A survey of badger setts and habitats within a total of 735 1km squares in the Irish Republic commenced in 1989, representing a sample of 1% of the country's land area. Over 36% of the survey is completed and this paper presents the initial findings: sampling effort has had a pronounced westerly bias to date and the findings do not reflect a statistically valid sample of the country. The overall density of active main setts is 0.57 km^{-2}, each representing a badger social group. There is evident regional variation, with lower densities in the west (Cos Galway and Mayo). Seventy-one per cent of all setts were active and the proportions of sett types were: 1 active main, 0.45 annexe, 0.95 subsidiary and 1.03 outlying. Main setts had, on average, 6.7 entrances. Forty-three per cent of setts occur in hedgerow or in treelines, 19% in woodland and 21% in scrub; 15.5% of setts have been disturbed, principally by blocking or digging.

INTRODUCTION

The badger (*Meles meles*) is widespread throughout Ireland (Ní Lamhna 1979) and is one of the most common of the larger mammals represented in the island's fauna. Fairley (1975) states that 'there is some reason for believing that the badger was scarcer in Ireland at the turn of the century than it is today, in the north especially'. There is little information available on the density and distribution of badgers in Ireland, and an equivalent body of information such as that compiled by the Mammal Society and many local badger groups throughout Britain (for a review see Cresswell *et al.* 1990) has not been available to zoologists in Ircland.

Since badgers may be implicated in the spread of infection with *Mycobacterium bovis* in cattle, there has been a need for a comprehensive survey of badgers in the country. The methodology adopted in this survey has been based largely on that of the Nature Conservancy Council's recent survey in Britain (Cresswell *et al.* 1990; Harris *et al.* 1989). This uniformity of approach allows for comparisons and conclusions to be drawn from results from the two islands with relative ease. In practice, the survey of badgers is achieved through a thorough enumeration of badger setts in a sizeable sample of 1km squares evenly distributed throughout Ireland. The aim is to sample 1% of the island's land area. A similar survey is being conducted in Northern Ireland in co-operation with Queen's University, Belfast (Feore *et al.* 1993).

The bulk of survey work commenced in January 1990, and has been undertaken principally by the staff of the National Parks and Wildlife Service, with a number of surveys being undertaken by volunteers. In addition to the enumeration of setts, habitats within the 1km squares are mapped, providing an opportunity to evaluate the relationship of badger densities to habitat variables. The survey data may be used to extrapolate to unsurveyed areas (through the inclusion of environmental variables), to examine relationships between badger densities and TB outbreaks in cattle, and as a baseline against which future changes in badger densities and habitats may be compared.

METHODS

A total of 735 1km squares was selected, each being the south-west corner of every 10km square in the country (total for the island 879; squares falling on the border are included within the survey for the Republic: it may not be feasible to survey all of these). Coastal squares that contain any land above tidal limits are included. Fifty field staff plus a number of volunteers have been involved in the survey, which is ongoing and due to be completed in spring 1992. Fieldworkers are provided with an enlarged 6":1 mile map (approximate scale 9":1 mile), with a superimposed grid representing 1ha squares, and five data sheets for recording habitat and environmental details and badger sett data. Habitats (listed in Appendix 2) are marked on the map using coloured pencils.

All areas of the 1km square are walked and searched for setts. The survey is carried out during a period of low vegetational cover: late October to mid-May. The size and activity of setts are noted (number of entrances and their level of use, presence of tracks, fresh digging, bedding, latrines etc.) and the setts are classified as main, annexe, subsidiary or outlier setts, in accordance with Thornton (1988). (See Appendix 1.) Habitat at the sett site is recorded, as is the soil type, presence of cattle in vicinity, and any evidence of disturbance to the sett. In the course of the field survey, surveyors are asked to mark in other signs of badgers observed, including hairs, latrines and footprints. Observations of other mammals are also recorded. Each square has taken an average of 2.1 man-days to complete.

Badgers are organised socially into territorial groups (Kruuk 1978), with a variable number of setts. A social group will contain, within its territory, one main sett (which is used throughout the year), an annexe sett (or setts) nearby (not always present), subsidiary setts often at an intermediate distance, and outlying setts often at a further distance. In Offaly, social groups used from one to eight setts over a two-year period, but in all cases no more than one active sett was identified per group (O'Corry-Crowe *et al.* 1993). The assessment of the number of social groups is dependent upon the correct classification of active main setts, which is best carried out after completion of a survey square; all records were checked after submission to the author, and re-evaluated, if necessary, in relation to sett size, activity and location with regard to other setts. Any queries regarding setts or habitat information were returned to the field staff for comments and/or revision.

RESULTS

To date, almost 40% of the survey in the Republic has been completed (266 squares). Survey effort has not been equally distributed and has a pronounced westerly bias (Fig. 1). The country has been divided into administrative districts (Fig. 1), and progress county by county is given in Table 1. The results presented within this paper are not statistically representative of the whole country and a statistical analysis of regional variation has therefore not been attempted prior to completion of the survey.

(b)

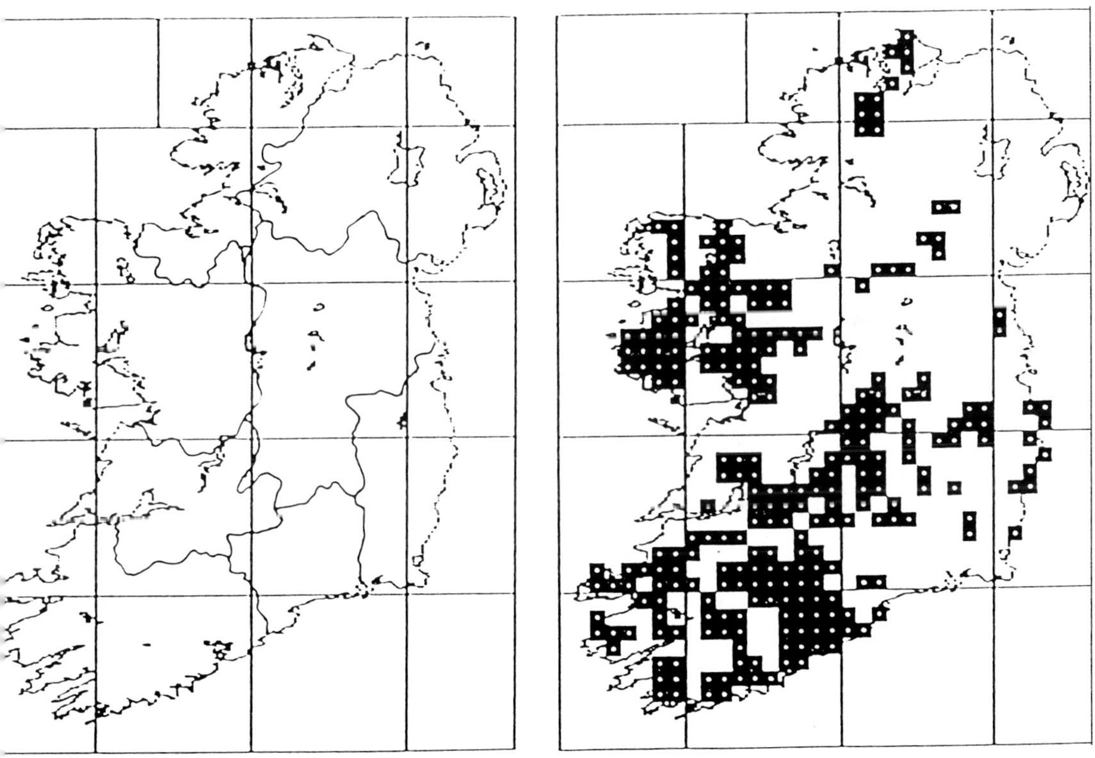

Fig. 1a. Administrative districts referred to in tables.

Fig. 1b. Distribution of survey effort to date: 10km squares blacked in represent the 1km squares within each of the 10km squares that have been completed.

Table 1. Number of survey squares given by county.

County	Total no. of survey squares	No. of squares surveyed	Percentage of squares surveyed
Carlow	8	1	12.5
Cavan	18	3	16.7
Clare	36	10	27.8
Cork	78	59	75.6
Donegal	56	11	19.6
Dublin	10	0	0
Galway	66	32	48.5
Kerry	54	28	51.9
Kildare	17	5	29.4
Kilkenny	20	3	15.0
Laois	19	6	31.6
Leitrim	16	2	12.5
Limerick	27	15	55.6
Longford	11	0	0
Louth	8	0	0
Mayo	16	31	50.0
Meath	28	2	7.1
Monaghan	13	4	30.8
Offaly	18	11	61.1
Roscommon	25	4	16.0
Sligo	19	0	0
Tipperary	43	23	53.5
Waterford	22	3	13.6
Westmeath	17	0	0
Wexford	25	5	20.0
Wicklow	19	8	42.1
TOTALS			
Republic	735	266	36.2
Northern Ireland	144	-	-
Ireland (total area)	879	-	-

The information on sett type and distribution is summarised in Table 2 for all setts, and in Table 3 for active setts only. The frequency of occurrence of setts within squares is shown in Fig. 2: many squares contain no setts at all, and a small proportion more than ten. However, no square surveyed to date has been found with more than three active main setts. The overall density of active main setts, each representing a badger social group, is 0.57km^{-2}. Initial indications are that densities fall into the range 0.5 to 0.9 social groups per 1km^2 over much of the country, a little lower in the south, and substantially lower in the west (Cos Galway and Mayo) at *c.* 0.3 social groups km^{-2}.

Table 2. Summary of badger sett data by district.

REGION	No. of squares surveyed	Total no. of setts observed	Mean no. of setts /km^2	TOTAL NO. OF SETT TYPES				MEAN NO. OF SETT TYPES/ KM^2			
				Main	Annexe	Subsid.	Outlier	Main	Annexe	Subsid.	Outlier
South-west	87	207	2.38	67	26	50	63	0.77	0.30	0.57	0.72
Mid-west	40	103	2.58	33	13	37	20	0.83	0.33	0.93	0.50
West	67	44	0.66	20	6	7	11	0.30	0.09	0.10	0.16
North-west	11	22	2.00	7	3	8	4	0.64	0.27	0.73	0.36
Midlands	28	91	3.25	30	8	26	27	1.07	0.29	0.93	0.96
South	14	43	3.07	8	6	15	14	0.57	0.43	1.07	1.00
East	19	60	3.16	19	7	17	17	1.00	0.37	0.89	0.89
Total area	226	570	2.14	184	69	160	156	0.69	0.26	0.60	0.59

Table 3. Summary of active badger sett data by district.

REGION	No. of squares surveyed	Total no. of active setts observed	Percentage of active setts/km^2	TOTAL NO. OF ACTIVE SETTS				MEAN NO. OF ACTIVE SETTS/KM^2			
				Main	Annexe	Subsid.	Outlier	Main	Annexe	Subsid.	Outlier
South-west	87	142	68.6	55	20	35	32	0.63	0.23	0.40	0.37
Mid-west	40	81	78.6	29	10	31	11	0.73	0.25	0.78	0.28
West	67	34	77.3	18	6	4	6	0.27	0.09	0.06	0.09
North-west	11	13	59.1	6	3	3	1	0.55	0.27	0.27	0.09
Midlands	28	71	78.0	24	6	19	22	0.86	0.21	0.68	0.79
South	14	31	72.1	6	4	9	12	0.43	0.29	0.64	0.86
East	19	60	53.3	14	7	9	2	0.74	0.37	0.47	0.11
Total area	226	404	70.9	152	56	110	86	0.57	0.21	0.41	0.32

(a)

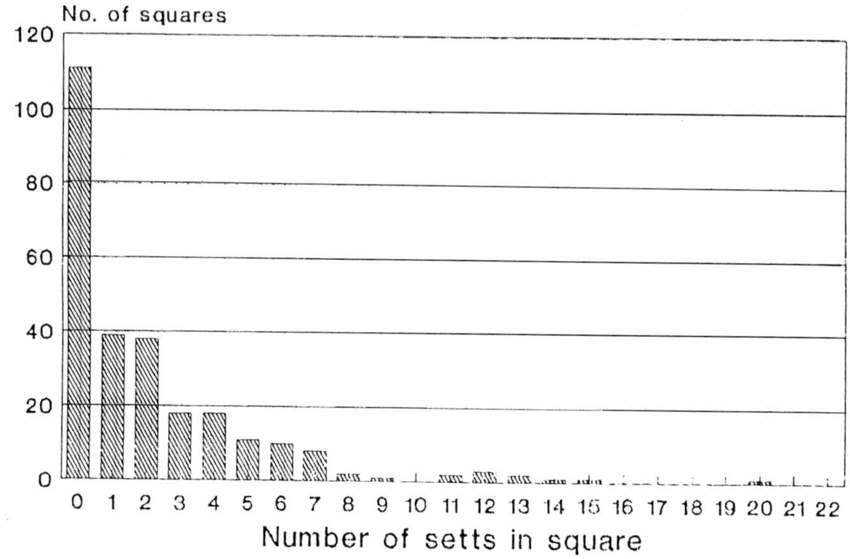

(b)

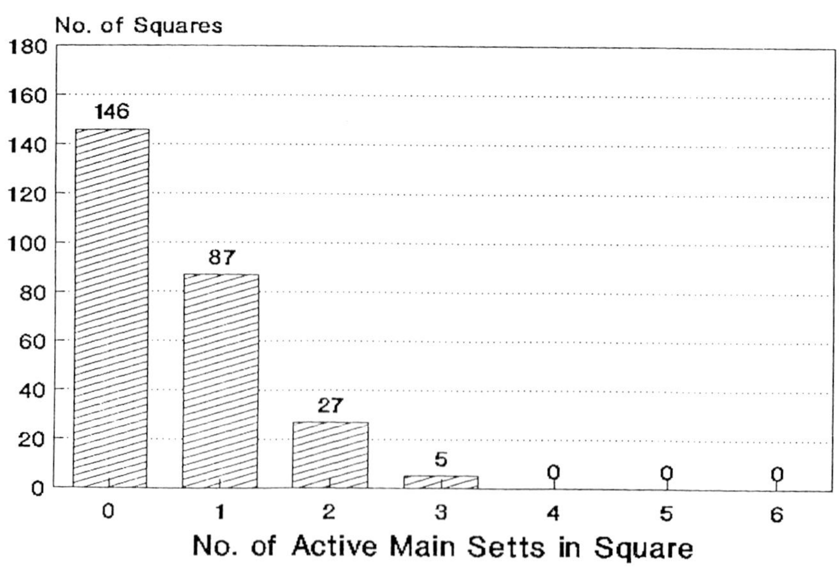

Fig. 2. Frequency distribution of densities (km^{-2}) of (a) all setts and (b) main setts.

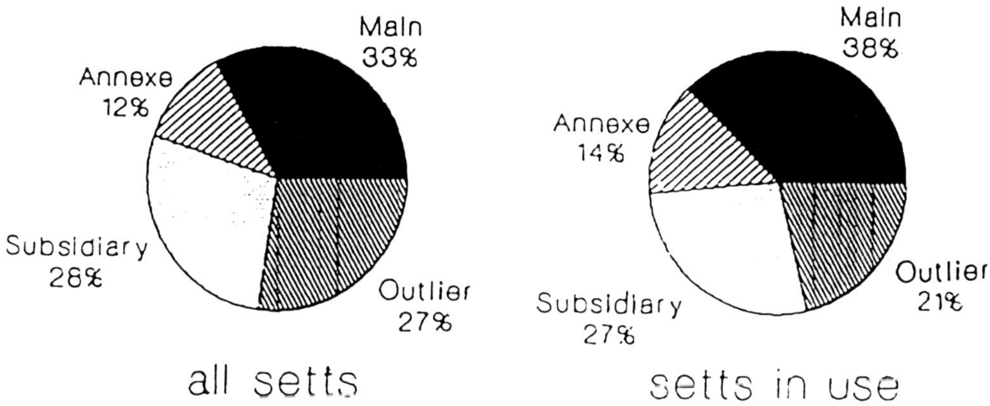

Fig. 3. Proportion of sett types present in squares surveyed.

Overall, the proportions of sett types were 1 active main sett, 0.45 annexe setts, 1.05 subsidiary setts, and 1.03 outlying setts (Fig. 3). Seventy-one per cent of all setts were recorded as active: 83% of main setts, 81% of annexe setts, 69% of subsidiary setts, and 55% of outlying setts. The largest main sett located within squares surveyed to date had 36 entrances, and the smallest 2 entrances. On average, main setts had 6.7 entrances (Table 4).

Figure 4 reveals that main setts with 4 or 5 entrances are the most frequent. Regional differences in size of main setts were not related to density but differences in main sett size do appear between districts, with means of *c.* 9 entrances for the south and north-west. Annexe setts have fewer entrances than main setts, averaging 4.1; subsidiaries average 2.7 entrances but may have as many as 8, and outlier setts have a single entrance.

Some entrances to an active sett are used more than others, and a number may reveal no active use. The overall proportions of well-used, partially used and disused entrances for each sett category are given in Table 5.

The habitats in which setts are located are summarised in Table 6. Setts are most commonly located within hedgerows (39%) and in low scrub (13.5%). Almost 43% of setts occur either in hedgerow or treelines, 19% in woodland (predominantly coniferous plantation), 21% in low or tall scrub and 10% in grassland. The remaining 7% of setts occur in other habitat categories. The proportion of outlier setts occurring in grassland is higher than that for the other sett types.

Levels of disturbance to badger setts average 15.5%, but 22.3% for main setts (Table 7). The majority of these instances relate to evidence of digging or blocking at the sett. It is not possible to evaluate a rate of disturbance as signs of digging or other interference may remain for some years after the event: some of the records of disturbance include setts known to have been blocked or dug up to six years previously.

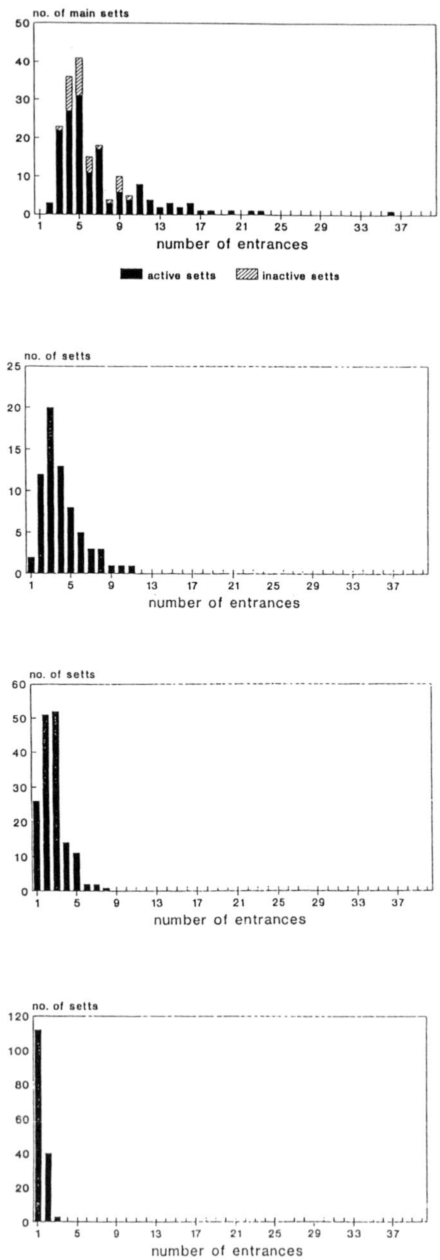

Fig. 4. Distribution of setts according to size as indicated by number of entrances: (a) main, (b) annexe, (c) subsidiary and (d) outlier.

Table 4. Size of setts given by number of entrances.

REGION	ALL SETTS				SETTS IN USE			
	Main	Annexe	Subsid.	Outlier	Main	Annexe	Subsid.	Outlier
South-west	7.1	4.3	2.5	1.2	7.4	4.6	2.6	1.1
Mid-west	5.6	4.2	2.4	1.3	5.8	3.9	2.4	1.2
West	6.2	4.8	2.7	1.6	6.3	4.8	3.5	1.7
North-west	9.3	3.3	2.8	1.8	10.3	3.3	2.7	2.0
Midlands	6.6	3.0	3.0	1.3	6.8	2.8	3.1	1.3
South	8.9	4.3	3.5	1.4	9.3	4.8	3.2	1.4
East	6.0	4.1	2.6	1.3	6.4	4.1	2.2	1.5
Mean	6.7	4.1	2.7	1.3	7.0	4.2	2.7	1.3
s.e.	0.32	0.25	0.10	0.04	0.38	0.29	0.13	0.05
Max. no. of entrances	36	11	8	3	36	11	8	3
Min. no. of entrances	2	1	1	1	2	1	1	1

Table 5. Use of entrances by badgers at the four sett categories given by mean number of entrances well used, partially used and disused.

ALL SETTS Active/Disused	MAIN	ANNEXE	SUBSID.	OUTLIER
Well-used entrances	3.4	1.9	1.0	0.4
Partially used entrances	1.0	0.5	0.4	0.2
Disused entrances	2.3	1.7	1.2	0.6
ACTIVE SETTS ONLY				
Well-used entrances	4.1	2.3	1.5	0.8
Partially used entrances	1.2	0.6	0.5	0.4
Disused entrances	1.7	1.3	0.7	0.1

Table 6. Location of setts according to habitat type.

CATEGORY						
Habitat description	All setts	% of all setts	Main	Annexe	Subsid.	Outlier
1 Hedgerow	220	38.9	50	26	76	68
2 Treelines	21	3.7	7	2	9	3
3 Semi-natural broad-leaved woodland	41	7.3	15	7	10	8
4 Broad-leaved plantation	16	2.8	6	3	4	3
5 Semi-natural coniferous woodland	1	0.2	1	0	0	0
6 Coniferous plantation	45	8.0	16	7	15	7
7 Semi-natural mixed woodland	3	0.5	1	0	0	2
8 Mixed plantation	1	0.2	0	1	0	0
9 Young/mixed broad-leaved plantation	0	0	0	0	0	0
10 Recently felled woodland	1	0.2	0	0	0	1
11 Parkland	0	0	0	0	0	0
12 Tall scrub	42	7.4	16	6	11	9
13 Low scrub	75	13.5	33	7	15	21
14 Bracken	10	1.8	4	2	4	0
15 Coastal sand-dunes	0	0	0	0	0	0
16 Coastal sand or mud-flats	0	0	0	0	0	0
17 Coastal shingle or boulder beaches	0	0	0	0	0	0
18 Lowland heaths	2	0.4	1	0	1	0
19 Heather moorlands	3	0.5	1	1	0	1
20 Blanket bog	1	0.2	0	0	0	1
21 Raised bog	1	0.2	1	0	0	0
22 Marginal inundations	0	0	0	0	0	0
23 Coastal marsh	0	0	0	0	0	0
24 Wet ground	1	0.2	0	0	0	1
25 Standing natural water	0	0	0	0	0	0
26 Standing man-made water	0	0	0	0	0	0
27 Running natural water	1	0.2	0	0	0	1
28 Running canalised water	3	0.5	1	1	1	0
29 Upland unimproved grassland	7	1.2	4	0	1	2
30 Lowland unimproved grassland	6	1.1	3	1	1	1
31 Semi-improved grassland	14	2.5	4	1	3	6
32 Improved grassland	29	5.1	8	2	5	14
33 Arable	10	1.8	3	0	2	5
34 Amenity grassland	0	0	0	0	0	0
35 Unquarried inland cliffs	1	0.2	1	0	0	0
36 Vertical coastal cliffs	0	0	0	0	0	0
37 Sloping coastal cliffs	0	0	0	0	0	0
38 Quarries and opencast mines	8	1.4	5	0	2	1
39 Bare ground	0	0	0	0	0	0
40 Built land	1	0.2	0	1	0	0
Totals	565	100	181	68	160	155

Table 7. Levels of disturbance to setts, including persecutive disturbance in the course of land development and legal removal of badgers.

	MAIN	ANNEXE	SUBSID.	OUTLIER	TOTAL
No. disturbed	41	9	25	13	88
Totals all setts	184	69	160	156	569
Percentage	22.3	13.0	15.6	8.3	15.5
Type of disturbance					
Digging	25	3	9	4	41
Blocking	6	3	10	4	23
Burning	1	0	2	1	4
Official removal	1	2	1	3	7
Snaring	1	0	0	0	1
Development	7	1	3	1	12

DISCUSSION

The National Badger Survey is based on a census of setts within 1% of the country's land area, and is almost 40% completed. Survey effort has a pronounced bias towards western counties at present, and extrapolations to the whole country or inferences with regard to regional variation are tentative and must be regarded as preliminary. There is also a need to calibrate the fieldwork using areas in which badger densities and sett use are known.

Nevertheless, there are clear indications of differences in badger ecology between Ireland and Britain. Densities of social groups are higher in Ireland: overall densities at 0.57 groups km^{-2} are higher than the mean reported for south-west England, 0.45 groups km^{-2} (Cresswell *et al.* 1990). The latter figure included urban areas. Substantial areas of south and south-west England, and also south and west Wales, supported badger densities in the range 0.5-0.7 badger groups km^{-2}. Densities of social groups reported here for the west of Ireland (*c.* 0.3 km^{-2}) are similar to those given by Cresswell *et al.* (1990) for Wales (0.29 km^{-2}), south-east England (0.32 km^{-2}) and the West Midlands (0.28 km^{-2}). Elsewhere in Britain, particularly in Scotland, badger densities are substantially lower.

Thornton (1988), following a survey of 42 tetrads (168km^{2}), reported main sett density for Cornwall and Devon of 0.58 km^{-2}. Density may be substantially higher, as on the Cotswold escarpment, Gloucestershire, where social group density was *c.* 3.8 km^{-2}

(Cheeseman, Cresswell *et al.* 1988). Densities there are the highest recorded for the species in Europe (Cheeseman, Jones *et al.* 1981).

In overall terms, badger group densities in Ireland are not, therefore, dissimilar to those in the areas of higher badger density in Britain, but large areas of very high badger density such as those in the Cotswolds are not known in Ireland. There is, however, a need to calibrate the fieldwork in this study using areas in which badger densities are known (O'Corry-Crowe *et al.* 1993; Sleeman and Mulcahy 1993).

The mean size of active main setts in Ireland (7.0 entrances) is substantially smaller than in Britain (11.9; Cresswell *et al.* 1990) and may indicate a smaller social group size in Ireland. Even in very low density areas (<0.1 groups km^{-2}), main setts in Britain averaged *c.* 8 entrances. Annexe, subsidiary and outlier setts were also, on average, larger than their counterparts in Ireland. Differences in the use of entrances indicate that, for any sett, the proportion of well-used entrances is higher in Ireland than in Britain. Thus for main setts 59% of entrances are well used in Ireland (cf. 48%), 56% of annexe entrances (cf. 34%), 56% of subsidiary sett entrances (cf. 24%), and 62% of outlier entrances (cf. 24%).

Despite these differences, the proportions of sett types were very similar to those in Britain, except for outliers (active main 1, disused main 0.21 (cf. 0.24 in Britain), annexe 0.45 (cf. 0.43), subsidiary 1.05 (cf. 0.86), outliers 1.03 (cf. 1.57); Cresswell *et al.* 1990). But in the higher density areas, the number of outlying setts (per active main sett) found in Britain was fewer (1.22). However, an intensive survey in the midlands revealed that the average size of each sett category was very similar to values obtained in Britain. Main setts had 11.91 entrances, annexes 5 entrances, subsidiary setts 3.9 entrances, and outliers 1.35 entrances (O'Corry-Crowe *et al.* 1993).

The substantially smaller size of main setts and the greater degree of use of the entrances in Ireland may result from limited opportunities for building large setts in Ireland: the area of woodland and number of woodland blocks is smaller, and hedgerow length believed to be greater, which may be the cause of a high proportion of setts being located in hedgerow and treeline in Ireland. In Britain, only 14% of active main setts were located in hedgerow (39% in Ireland), 50% in woodland (cf. 19%), and 7% in scrub (cf. 21%). Hedgerows do not, generally, allow the construction of large setts, and large setts in hedgerow are also more liable to destruction or disturbance from general farming activity in their proximity. Nevertheless, in an intensively studied area in Co. Offaly, over 50% of setts occurred in hedgerows which occupied less than 4% of the area (O'Corry-Crowe *et al.* 1993).

It is not clear whether group size is reduced with smaller but 'more active' setts. Results from the midlands suggest group sizes of *c.* 5.8 (O'Corry-Crowe *et al.* 1993), similar to the mean of 5.9 adults arrived at by Cresswell *et al.* (1990) in a review of published data on social group size in Britain. Further information on social group size in Ireland is essential, in view of the substantial differences in sett size and activity reported here: group size averaged over the country, including lower density areas, is likely to be lower than 5.8.

A tentative extrapolation of the current data to the whole island would suggest that there are *c.* 50,000 social groups present, comprising *c.* 250,000 adult badgers (using a mean social group size of 5.0). This is higher than the badger population in Britain (*c.* 43,000 social groups; Cresswell *et al.* 1990), but that population is largely concentrated in the south and south-west, in perhaps a similar land area.

Levels of disturbance to badger setts appear to be of a similar order to those given by Cresswell *et al.* (British data: 15.7% of main setts with holes blocked, 10.5% dug, 1.0% snared), but digging is more common than blocking in Ireland.

ACKNOWLEDGEMENTS

Results presented here are based on the fieldwork of the field staff of the National Parks and Wildlife Service and a number of volunteers. The assistance of Penny Cresswell in the initial stages is gratefully acknowledged. The maps were prepared with the aid of a program written by Des Higgins.

REFERENCES

CHEESEMAN, C.L., CRESSWELL, W.J., HARRIS, S. and MALLINSON, P.J. 1988 Comparison of dispersal and other movements in two badger (*Meles meles*) populations. *Mammal Review* **8**, 51-9.

CHEESEMAN, C.L., JONES, G.W., GALLAGHER, J. and MALLINSON, P.J. 1981 The population structure, density and prevalence of tuberculosis (*Mycobacterium bovis*) in badgers (*Meles meles*) from four areas of south-west England. *Journal of Applied Ecology* **18**, 795-804.

CRESSWELL, P., HARRIS, S. and JEFFRIES, D.J. 1990 *The history, distribution, status and habitat requirements of the badger in Britain.* Peterborough. Nature Conservancy Council.

FAIRLEY, J. S. 1975 *An Irish beast book.* Belfast. Blackstaff Press.

FEORE, S., SMAL, C.M. and MONTGOMERY, W.I. 1993 Survey of badger setts in Northern Ireland: progress report. In T. J. Hayden (ed.), *The badger,* 23-5. Dublin. Royal Irish Academy.

HARRIS, S., CRESSWELL, P. and JEFFRIES, D. 1989 *Surveying badgers.* Mammal Society Occasional Publication No. 9.

KRUUK, H. 1978 Spatial organization and territorial behaviour of the European badger (*Meles meles*). *Journal of Zoology, London* **184**, 1-19.

NÍ LAMHNA, E. 1979 *Provisional distribution atlas of amphibians, reptiles and mammals in Ireland.* Dublin. An Foras Forbartha.

O'CORRY-CROWE, G., EVES, J. and HAYDEN, T.J., 1993 Sett distrbution, territory size and population density of badgers (*Meles meles* L.) in east Offaly. In T.J. Hayden (ed.), *The badger*, 35-56. Dublin. Royal Irish Academy.

SLEEMAN, D.P. and MULCAHY, M.F 1993 Behaviour of Irish badgers in relation to bovine tuberculosis. In T.J. Hayden (ed.), *The badger,* 154-65. Dublin. Royal Irish Academy.

THORNTON, P.S. 1988 Density and distribution of badgers in southwest England: a predictive model. *Mammal Review* **18**, 11-23.

APPENDIX 1

Main setts: These usually have a large number of entrances (used and disused) with conspicuous soil heaps. The setts look well used, with the paths between entrances and to and from the sett being obvious and well worn. Main setts are breeding setts and are normally in continuous use. However, main setts may become disused owing to disturbance or some other reason, and should then be recorded as disused main setts.

Annexe setts: These are close to a main sett, between 50m and 150m away, and are usually connected to the main sett by well-worn paths. They usually have several holes, but may not be in use all the time, even if the main sett is very active.

Subsidiary setts: These have an intermediate number of entrances and are not connected to another sett by obvious paths. They are usually at least 50m from a main sett and are not continuously active.

Outlier setts: These usually have only one or two holes, often with little spoil outside the hole, and have no obvious path connecting with another sett. They are used only sporadically, and, when not in use by badgers, they may be taken over by foxes and even by rabbits.

In: Hayden, T.J. (ed.) 1993 *The badger*, 23-25. Dublin. Royal Irish Academy.

SURVEY OF BADGER SETTS IN NORTHERN IRELAND: PROGRESS REPORT

S. Feore,* C.M. Smal and W.I. Montgomery*†**

*** School of Biology and Biochemistry, Queen's University, Belfast.**
**** Wildlife Service, Office of Public Works, Bray, Co. Wicklow.**

†To whom correspondence should be addressed.

INTRODUCTION

The distribution and abundance of the badger in Northern Ireland are poorly known. In Britain, however, numerous surveys have been carried out on both a national and a local scale. In 1989, a Nature Council/Mammal Society nationwide badger survey was completed by Cresswell *et al.* (1990). In their survey, badger densities were estimated in relation to environmental and habitat variables in over 1% of the land area in order to develop predictive models allowing extrapolation to unsurveyed areas.

In October 1989, a nationwide survey of badgers in Ireland was planned to complement the recent British survey using the same methods (Smal 1993). In October 1990, the project began at Queen's University, Belfast, and it was agreed that the survey of Northern Ireland be carried as part of the overall Irish survey so that comparative data would be available for the whole of the British Isles.

Reported links between bovine tuberculosis in badgers and cattle in Britain have made badgers the subject of epidemiological studies. The high incidence of tuberculosis in cattle in the Republic of Ireland resulted in limited badger control operations there. However, the routes of infection remain unclear and the ethical, conservation and economical concerns are considerable. This survey of badgers is therefore very important and will establish the status of the badgers, providing baseline data which will allow the impact of badger removal to be assessed, as well as the impact of other threats such as badger persecution, road deaths and habitat loss.

The survey aims to:

1. assess the overall presence/absence of badgers in Ireland in 10km squares,
2. sample a minimum of 1% of the total land area,
3. relate density to topography, geology and soils, climate, land use, vegetation type and develop a predictive model allowing estimation of badger densities in areas not surveyed,
4. validate or refine British models in the Irish context, effecting comparison of Irish data with those available for Britain,
5. investigate the incidence of tuberculosis in cattle in relation to farming practice and badger densities and to attempt to isolate confounding effects,
6. assess the level of human disturbance to badgers, and
7. provide a reliable baseline for future badger surveys.

METHODS

1. A 1km square is being sampled within each 10km square on the Irish national grid.
2. The survey within each 1km square consists of locating and counting every badger hole.
3. All setts are classified according to accepted definitions (Cresswell *et al.* 1990). Main sett, activity, and badger signs are recorded for each sett as well as the presence of foxes, rabbits and human disturbance.
4. Detail of the habitat is indicated on an enlarged scale map and assigned a habitat category including details of land use. The landscape is also placed within a land class.
5. Other mammals are noted for a survey of mammals based on presence within each 1km square.

RESULTS AND DISCUSSION

Since the survey began in November 1990, 102 of the 144 km-square areas in Northern Ireland have been surveyed as of April 1992. The remaining areas are due to be completed by May 1992.

The survey will provide data on sett density. The density of active main setts is indicative of the density of badger social groups. Table 1 provides a summary of the data received since March 1991. It shows the results from the available data for the North and the Republic up to April 1992.

Table 1. The active setts observed for the Republic and the North of Ireland during the survey.

REGION	No. of squares surveyed	Total no. of active setts observed	Percentage of setts active	TOTALS OF SETTS ACTIVE				MAIN SETTS/KM^2			
				Main	Annexe	Subsid.	Outlier	Main	Annexe	Subsid.	Outlier
South-west	128	184	68.7	74	27	45	38	0.58	0.21	0.35	0.30
Mid-west	84	180	93.3	59	29	66	26	0.70	0.35	0.79	0.31
West	130	78	123.8	37	18	13	10	0.28	0.14	0.10	0.08
North-west	40	22	88.0	13	4	4	1	0.33	0.10	0.10	0.03
Midlands	86	159	103.9	56	15	40	48	0.65	0.17	0.47	0.56
South	51	85	101.2	32	7	25	21	0.63	0.14	0.49	0.41
East	48	65	79.3	31	12	12	10	0.65	0.25	0.25	0.21
Republic	567	773	89.1	302	112	205	154	0.53	0.20	0.36	0.27
North	91	188	69.1	68	35	50	35	0.75	0.38	0.55	0.38

The data collected from the sett survey in conjunction with work being carried out on establishing social group size, along with the recently completed land classification of Northern Ireland and other environmental data, will all be analysed to provide an estimate of badger density and total population size for Northern Ireland.

REFERENCES

CRESSWELL, P., HARRIS, S. and JEFFERIES, D. 1990 *The history, distribution, status and habitat requirements of the badger in Britain.* Peterborough. Nature Conservancy Council.

SMAL, C. 1993 The National Badger Survey: preliminary results for the Irish Republic. In T.J. Hayden (ed.), *The badger,* 9-22. Dublin. Royal Irish Academy.

In: Hayden, T.J. (ed.) 1993 *The badger*, 26-34. Dublin. Royal Irish Academy.

BADGER SETTS AS A LIMITING RESOURCE

T.J. Roper

School of Biological Sciences, University of Sussex, Brighton

ABSTRACT

The European badger's habit of constructing extensive underground burrows or 'setts' is one of its most striking characteristics, yet little thought has been given to the question of why badgers need setts or whether the availability of sett sites affects the distribution and abundance of badgers.

Anecdotal evidence suggests that (a) badgers are extremely reluctant to abandon an established sett, (b) setts can remain in occupation for centuries, and (c) setts can become very large. More systematic results from sett excavations confirm the latter point, showing that setts can contain hundreds of metres of underground tunnels and dozens of chambers, nests and entrances. I argue, from this evidence, that the sett is a valuable resource for the animals possessing it: something that has taken time and energy to acquire and cannot easily be replaced. I review various hypotheses as to why badgers might benefit from having such extensive setts and suggest that well-established setts may increase breeding success. I also suggest that sociality has evolved in badgers at least in part because offspring stand to gain more from remaining in the natal group and inheriting the parental sett than from leaving and trying to construct new setts of their own.

In addition, I review evidence suggesting a correlation between the number of setts in an area and the type of soil. Whereas previous researchers have suggested that badger population density is determined largely by the pattern of food availability, I suggest that the availability of suitable sett sites is at least as important a factor.

INTRODUCTION

The European badger's habit of constructing and utilising extensive underground burrows, or 'setts', is one of its most striking characteristics. Badgers are unique amongst carnivores in the degree to which they are anatomically and physiologically adapted to a semi-fossorial existence (Neal 1977; Long and Killingley 1983), in the sheer scale and complexity of their subterranean excavations (Neal and Roper 1991), and in the fact that they spend more than half their lives underground (e.g. Jensen 1959).

Despite this, very little thought has been given to the question of why badgers need elaborate setts, or whether the availability of suitable sett sites affects the distribution and abundance of badgers. In this paper I describe what is known about the architecture of setts and review various hypotheses concerning sett function. I also discuss factors which affect the siting of setts and consider the importance of sett availability from the point of view of badger population density. My overall theme is that the sett is a vital resource, the significance of which cannot be ignored if we are fully to understand badger behaviour and ecology.

I consider only 'main setts' (*sensu* Neal 1948; 1977; Kruuk 1978) since these are larger than other types of sett and are presumably more important in terms of survival value. Information about 'outliers', 'subsidiary setts' and 'annexes' is provided by Neal and Roper (1991) and Roper *et al.* (1991 a, b).

STRUCTURE AND FUNCTION OF MAIN SETTS

Main sett age, size and architecture as evidence of resource value

Many anecdotal reports suggest that badgers are reluctant to leave a well-established sett. For example, Wijngaarden and Peppel (1964) describe how badgers continued to inhabit a sett that was partially destroyed by excavation of a marl-pit, despite the fact that the badgers had to scramble up the walls of the pit in order to gain access to what was left of their home. Similarly, Roper *et al.* (1991a) report great difficulty in excluding badgers from a sett that was in the path of a road development scheme.

The fondness of badgers for established setts is also suggested by evidence that setts can remain in use for centuries. Neal (1977), for example, describes a sett in Hertfordshire that was well established at the end of the eighteenth century and was still inhabited almost 200 years later. Ancient place-names that include the word 'badger' or some variant of it often prove to refer to localities where active setts still exist (e.g. Wijngaarden and Peppel 1964; Long and Killingley 1983). Thus, the sett is a traditional home, handed on within a social group from generation to generation.

The fact that badgers rarely desert an established sett is one reason for thinking that setts are a valuable resource; another is the great size that such setts can attain. Wijngaarden and Peppel (1964) refer to a sett with over 100 entrances, covering an area exceeding 1ha, while Neal (1948) describes one whose entrances were scattered over an area of more than 0.5ha. Thompson (1931) tells how a terrier was sent into a sett in the Cotswolds and eventually emerged from what had been thought to be another sett, 100m away and on the other side of the hill. In my own study area in East Sussex, setts with more than 50 entrances, covering areas of a few hundred m^2, are not uncommon (Roper, unpublished observation).

Additional and more precise evidence concerning sett size is provided by systematic excavation, measuring and mapping of main setts. I have found descriptions of seven such excavations in the literature, five of which included a detailed plan of the sett in question. Table 1 summarises information about these seven setts and Fig. 1 shows plan views of two of them. In two cases (setts 4 and 7) excavation was not complete, but enough of the sett was revealed to give a good idea of its overall size and structure.

Table 1 and Fig. 1 show that while main setts differ considerably in size they can be very large indeed; and even relatively small setts contain a tunnel network of surprising length and complexity. The smallest sett, despite having an estimated internal volume of only 1.9m^3, nevertheless contained 43m of tunnels; while the largest, with an internal volume of 39m^3, contained a tunnel network totalling almost 1km in length, together with dozens of underground chambers. Estimates of the weight of soil removed during construction of the seven setts (based on a soil density of 1.6 tonnes/m^3: Leeson and Mills 1977) vary from 3 to 62 tonnes. Clearly a large sett can represent the investment, by successive generations of badgers, of enormous amounts of time and energy.

Table 1. Dimensions and contents of seven excavated main setts. Sources: 1-3 Likhachev (1956), 4 Frewin (1976), 5-6 Leeson and Mills (1977), 7 Roper *et al.* (1991b).

Sett no.	Area (m^2)	Volume (m^3)	Soil equiv.	Tunnel length	Entrance	Chambers	Bedding (tonnes)
1	46	1.9	3.0	43	4	1	-
2	54	3.1	5.0	39	2	3	-
3	130	4.7	7.5	81	3	1	-
4	304	4.6	7.4	102	2	8	8
5	525	15.3	24.5	310	15	39	26
6	740	25.2	40.3	360	38	78	15
7	1744	38.7	61.9	879	198	50	28

Sett function

Why do badgers need setts with multiple entrances, complex networks of underground tunnels and, in most cases, a choice of chambers and nests? One possible answer is that since the size of a sett is largely a function of its age (Neal and Roper 1991) and since successful breeding does sometimes occur in setts with a single entrance (Neal 1977), large setts need have no specific survival value: they are simply a consequence of the badger's 'insatiable need for excavating' (Wijngaarden and Peppel 1964, 17). But digging is an activity that consumes time and energy, so a tendency to go on extending setts indefinitely would surely have disappeared in the course of evolution if it had no survival value. Furthermore, the fact that breeding is possible in small setts does not mean that there is no *additional* fitness to be gained from inhabiting a large sett.

Several authorities (Likhachev 1956; Neal and Harrison 1958; Wijngaarden and Peppel 1964) state that successful breeding only occurs in setts that are several years old, suggesting that larger setts are more suitable for the rearing of cubs. There are several possible reasons why the physical environment offered by a large sett might favour improved cub survival. For example, cubs are more vulnerable than adults to predation, and a large sett may offer better protection. They are more susceptible to cold and damp, and, being confined to the sett for at least the first two months of life, they may require especially good conditions of ventilation.

Another factor affecting productivity, and possibly related to sett size, is competition amongst the female members of a social group for the opportunity to breed. In areas of high population density, badger groups generally contain more than one adult female (e.g. Kruuk 1978; Cheeseman *et al.* 1981), yet it is common for only one of these females to produce cubs in any one year (Cheeseman *et al.* 1987; Kruuk 1989). Furthermore, the probability that a female will breed successfully increases with age (Cheeseman *et al.* 1987). These facts suggest a dominance effect amongst the female members of a social group, whereby an older individual prevents younger subordinates from breeding successfully, perhaps by physically harassing them and/or their cubs (Kruuk 1989; Lüps

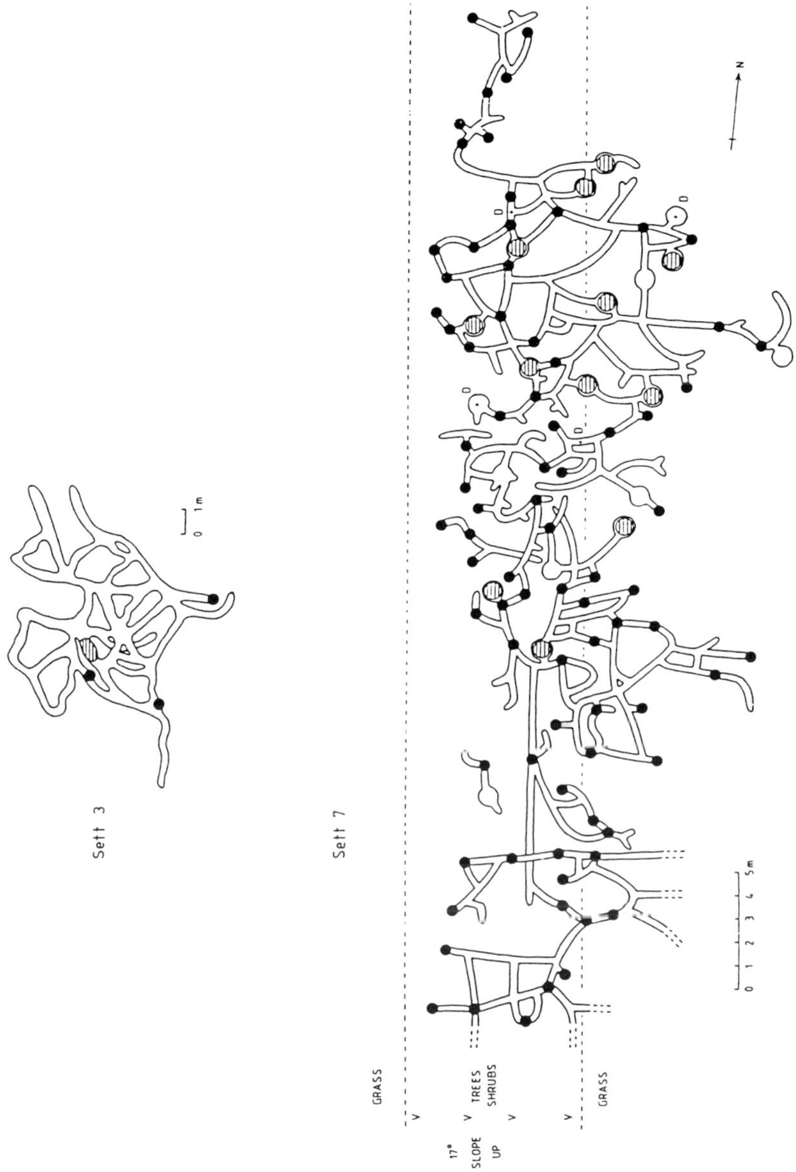

Fig. 1. Plan views of two badger setts. (a) Sett no. 3 in Table 1, redrawn from Likhachev (1956). (b) Sett no. 7 in Table 1, redrawn from Roper *et al.* (1991b) by permission of the Zoological Society of London. This sett measured 109m from the southern end to the northern end: only the first 44m of the northern end are shown.

and Roper 1990). If such a mechanism of within-group reproductive suppression does occur, a larger sett might allow a subordinate female to escape from the influence of a dominant one by using a different part of the sett. Thus, larger setts might produce more cubs by allowing females more breeding space.

A number of other advantages, not associated with breeding, can be suggested for large setts. First, badgers and their nests are usually infested with ectoparasites (Hancox 1980; 1988; Roper *et al.* 1991b), so it may be advantageous for individuals to be able to move from one nest to another when parasite levels build up. Second, main setts may offer a warmer and/or safer environment for overwintering (Kruuk 1989). And third, large complex setts with many tunnels and entrances may offer a better escape from, or defence against, predators. At present, unfortunately, there is little to choose between these and other hypotheses because we know almost nothing about the internal environment that setts offer (see Neal and Roper 1991) or about how the underground space within a sett is used by the different members of a badger group (Roper and Christian, in press).

The sett as a resource favouring sociality

Whatever the proximate reason why large setts are adaptive, the idea that the sett is a significant resource, with considerable survival value, may be relevant to the issue of why badgers are social. As Kruuk (1989) points out, none of the factors conventionally cited to explain carnivore sociality seem to apply to badgers: badgers do not cooperate in hunting, rearing offspring or defending themselves against predators. Kruuk suggests instead that sociality in badgers has arisen because the primary food of badgers (earthworms) is distributed in such a way that the smallest territory that will support a single pair of animals will support additional group members without there being additional competition for food. Thus sociality has arisen not because of some special advantage favouring it, but because of its lack of disadvantages in the form of competition for food (Kruuk 1986; 1989).

I wish to argue that in explaining sociality we might do better to focus on the sett rather than on food as the critical resource. Setts after all are used communally, whereas the badger's food supply is not: the individual members of a badger group forage solitarily (Kruuk 1978; Shepherdson *et al.* 1990). If it is really true that a well-established sett improves the chance of successful breeding or overwintering, and if such a sett takes several years to dig, starting from scratch, this could explain why offspring remain within the group rather than dispersing. In other words, it might make more sense for a young badger to remain within its natal group in the hope eventually of inheriting the family sett rather than leaving and trying to construct a new sett of its own. What we need to know, in order to evaluate this hypothesis, is what benefit, in terms of lifetime reproductive success, is conferred on the inhabitants of setts of different sizes and ages.

THE DISTRIBUTION OF SETTS

Factors affecting sett density

While badgers are widespread throughout Europe and much of Asia, sett density varies considerably from place to place, both at a national and a local level (see review, Neal and Roper 1991). What is the cause of this variability?

Neal and Roper (1991) review a number of studies suggesting a strong correlation between sett density and two habitat variables: soil type and the presence of tree cover. For example, in the most recent and most detailed of such studies Thornton (1988) surveyed about 100 setts in Devon and Cornwall and scored for each of them 18 separate land-use and habitat variables. She then subjected the data to multivariate analysis to determine which variables best predicted sett density. The results showed that the only variables to correlate significantly with sett density were ones relating to soil diggability, hilliness and availability of tree cover, i.e. variables that reflect the suitability of an area for sett digging. Variables related to food availability did *not* correlate with sett density, a point to which I return below.

While the idea that sett density is strongly correlated with soil type at a local level is not new (see, for example, Dunwell and Killingley 1969; Kruuk 1978; Bock 1986; 1988), it is less generally realised that the same correlation is obtained on a larger scale. E.D. Clements, chief recorder of badger setts for the Mammal Society, has kindly allowed me to see his records of the locations of the 1088 main setts known to exist in an area of about 10,000km^2 in the south-east of England. When the setts are plotted on a geological map a striking correlation emerges: 87% of setts are found in areas of chalk or Hastings beds, which account for only 55% of the total area, whereas only 13% of setts are found on clay soils, which account for 45% of the total area. The resultant association between sett density and soil type is highly significant statistically (Table 2). Furthermore, as Clements (pers. comm.) points out, the correlation is even more impressive in reality than appears from the data, because most of the setts recorded as being in clay are in fact in sandy outcrops too small to appear on the geological map. In short, badgers are notably absent from areas where the soil is not well drained and easily dug.

Table 2. Relationship between sett density and soil type for a sample of 1088 main setts in a region of about 10,000km^2 in south-east England (Clements, unpublished data, 1987). $X^2 = 452.8$, df = 4, $p < 0.001$.

Soil type	% area	Actual no. of setts	Expected no. of setts
Hastings beds	30.8	497	335
Chalk	24.3	448	264
Gault clay/greensand	18.0	70	196
Weald clay	23.8	63	259
London clay	3.1	10	34

Sett density and badger population density

It is well established that badgers are socially territorial: that is, each territory contains a single social group and each social group utilises a single main sett (e.g. Neal 1977; Kruuk 1978; 1989; Cheeseman 1979; Roper *et al.* 1986). Furthermore, although there are considerable differences in territory size between different regions, average social group size remains relatively invariant at about six adults per group (for references see

Clements *et al.* 1988). It follows that badger population density (number of groups x group size) is positively correlated with sett density and negatively correlated with territory size.

If we are to understand badger population dynamics and ecology, it is important to know how these three variables (population density, sett density and territory size) are causally related (Neal and Roper 1991). The most widely accepted hypothesis, proposed by Kruuk and others (Kruuk 1986; Kruuk and Macdonald 1985), is that territories are optimally sized so as to provide sufficient food to support a social group. According to this idea, food availability dictates territory size, which in turn dictates the number of social groups, and hence the number of setts, per unit area (see Fig. 2(a)). But as we have seen, such evidence as is currently available shows no correlation between sett density and habitat variables that would be expected to indicate food availability (Thornton 1988; see also Brown 1983). Rather, sett density is dictated by the suitability of the soil for sett construction.

I wish to argue that availability of suitable sett sites (which is a function of factors such as soil type, hilliness and tree cover) determines sett density directly, and thereby dictates social group density. Social group density, I suggest, then determines territory size: that is, territory size is primarily a function of intruder pressure (the number of groups competing for space in a given area) rather than of food availability (see Fig. 2,(b)). Of course, food availability must ultimately put a lower limit on territory size, because a territory that is too small to feed its occupants is unviable. However, there is no evidence that this limit has been reached, even in areas of high population density such as occur in the south of England. In such areas territories can be as small as 30ha (Cheeseman *et al.* 1981; Roper *et al.* 1986), yet breeding continues at a normal rate (Cheeseman *et al.* 1987).

(a)

Food density --------> Territory size -----> Social group density -----> Sett density

(b)

Sett availability ----> Sett density ----> Social group density ----> Territory size

Fig. 2. Two alternative hypotheses concerning the factors determining sett density, territory size and social group density.

CONCLUSIONS

Much of the work that has been done on badgers in the last two decades has focused on the relationship between food supply and variables such as group size, territory size and population density. I have argued, in this paper, that sett availability can just as plausibly fill the roles previously assigned to food availability: that is, it can act not only as a proximate determinant of group size, territory size and population density but also as an ultimate reason for the evolution of sociality and territoriality in badgers. The point that I most wish to emphasise, however, is that in order to survive and reproduce, badgers need a secure, dry sett just as much as they need food. If we are to understand badgers fully we need to consider all the resources that are vital to them, and these include the availability of suitable sett sites.

ACKNOWLEDGEMENTS

The author was supported by a grant from the UK Science and Engineering Research Council.

REFERENCES

BOCK, W.F. 1986 Kriterien zur Grossenbeurteilung der Baue des Dachses *Meles meles*. *Säugetierkundliche Mitteilungen* **33**, 227-34.

BOCK, W.F. 1988 Die Bedeutung des Untergrundes für die Grosse von Bauen des Dachses (*Meles meles*) am Beispiel zweier Gebiete Sudostbayerns. *Zeitschrift für Säugetierkunde* **53**, 49-57.

BROWN, C.A.J. 1983 Prey abundance of the European badger, *Meles meles* L., in north-east Scotland. *Mammalia* **47**, 81-6.

CHEESEMAN, C.L. 1979 The behaviour of badgers. *Applied Animal Ethology* **5**, 193-8.

CHEESEMAN, C.L., JONES, G.W., GALLAGHER, J. and MALLINSON, P.J. 1981 The population structure, density and prevalence of tuberculosis (*Mycobacterium bovis*) in badgers (*Meles meles*) from four areas in south-west England. *Journal of Applied Ecology* **18**, 795-804.

CHEESEMAN, C.L., WILESMITH, J.W., RYAN, J. and MALLINSON, P.J. 1987 Badger population dynamics in a high-density area. *Symposia of the Zoological Society of London* **58**, 279-94.

CLEMENTS, E.D., NEAL, E.G. and YALDEN, D.W. 1988 The National Badger Sett Survey. *Mammal Review* **18**, 1-9.

DUNWELL, M.R. and KILLINGLEY, A. 1969 The distribution of badger setts in relation to the geology of the Chilterns. *Journal of Zoology, London* **158**, 204-8.

FREWIN, B.C. 1976 The excavation of badger setts at Stantonbury and Milton Keynes village. *Milton Keynes Natural History Society* **2**, 20-8.

HANCOX, M. 1980 Parasites and infectious diseases of the European badger (*Meles meles* L.): a review. *Mammal Review* **10**, 151-62.

HANCOX, M. 1988 The nidicolous fauna of badger setts. *Entomologists' Monthly Magazine* **124**, 93-5.

JENSEN, P.V. 1959 Lidt om graevlingen. *Naturens Verden* **11**, 289-320.

KRUUK, H. 1978 Spatial organisation and territorial behaviour of the European badger *Meles meles*. *Journal of Zoology, London* **184**, 1-19.

KRUUK, H. 1986 Dispersion of badgers *Meles meles* (L., 1758) and their resources: a summary. *Lutra* **29**, 12-15.

KRUUK, H. 1989 *The social badger*. Oxford University Press.

KRUUK, H. and MACDONALD, D.W. 1985 Group territories of carnivores: empires and enclaves. In R.M. Sibly and R.H. Smith (eds), *Behavioural ecology: the ecological consequences of adaptive behaviour,* 521-36. Oxford. Blackwell.

LEESON, R.C. and MILLS, B.M.C. 1977 *Survey of excavated badger setts in the county of Avon.* MAFF ADAS Report, April 1977.

LIKHACHEV, G.N. 1956 Some ecological traits of the badger of the Tula Abatis broad-leaved forests. In P.B. Jurgensen (ed.), *Studies of mammals in government preserves,* 72-94. Moscow. Ministry of Agriculture USSR.

LONG, C.A. and KILLINGLEY, C.A. 1983 *The badgers of the world.* Springfield, Illinois. Charles C. Thomas.

LÜPS, P. and ROPER, T.J. 1990 Cannibalism in a female badger (*Meles meles*): infanticide or predation? *Journal of Zoology, London* **221**, 314-15.

NEAL, E. 1948 *The badger.* London. Collins.

NEAL, E. 1977 *Badgers.* Poole, Dorset. Blandford.

NEAL, E.G. and HARRISON, R.J. 1958 Reproduction in the European badger (*Meles meles* L). *Transactions of the Zoological Society of London* **29**, 67-131.

NEAL, E.G. and ROPER, T.J. 1991 The environmental impact of badgers (*Meles meles* L.) and their setts. In P.S. Meadows and A. Tufail (eds), *The environmental impact of burrowing animals and animal burrows,* 89-106. London. Zoological Society of London.

ROPER, T.J. and CHRISTIAN, S.F. 1992 Sett use in badgers *Meles meles.* In I.G. Priede and S.M. Swift (eds), *Radio telemetry: remote monitoring and tracking of animals,* 661-9. Chichester. Ellis Harwood.

ROPER, T.J., SHEPHERDSON, D.J. and DAVIES, J.M. 1986 Scentmarking with faeces and anal secretion in the European badger. *Behaviour* **97**, 94-117.

ROPER, T.J., TAIT, A.I., CHRISTIAN, S. and FEE, D. 1991a Excavation of three badger (*Meles meles* L.) setts. *Zeitschrift für Säugetierkunde* **56**, 129-34.

ROPER, T.J., TAIT, A.I., FEE, D. and CHRISTIAN, S. 1991b Internal structure and contents of three badger (*Meles meles* L.) setts. *Journal of Zoology, London* **225**, 115-24.

SHEPHERDSON, D.J., ROPER, T.J. and LÜPS, P. 1990 Diet, food availability and foraging behaviour of badgers (*Meles meles* L.) in southern England. *Zeitschrift für Säugetierkunde* **55**, 81-93.

THORNTON, P.S. 1988 Density and distribution of badgers in south-west England: a predictive model. *Mammal Review* **18**, 11-23.

THOMPSON, A.R. 1931 *Nature by night.* London. Ivor Nicholson and Watson.

WIJNGAARDEN, A. VAN and PEPPEL, J. VAN DE 1964 The badger, *Meles meles* L., in the Netherlands. *Lutra* **6**, 1-60.

In: Hayden, T.J. (ed.) 1993 *The badger*, 35-56. Dublin. Royal Irish Academy.

SETT DISTRIBUTION, TERRITORY SIZE AND POPULATION DENSITY OF BADGERS (*MELES MELES* L.) IN EAST OFFALY

G. O'Corry-Crowe,* J. Eves and T.J. Hayden*†**

* **Mammal Research Group, Department of Zoology, University College, Dublin.**
** **District Veterinary Office, Department of Agriculture and Food, Tullamore, Co. Offaly.**

†To whom correspondence should be addressed.

ABSTRACT

Although the badger, *Meles meles* (L.,1758), is a common and widespread member of the Irish mammalian fauna, precise information on its distribution and population density is only now becoming available. A detailed study of an area (16km^2) in east Offaly has been carried out to determine what factors influence sett density, territory size, social group size and population density, and to provide information which may aid in the calibration of the National Badger Survey. Furthermore, the effect of removal of badgers on the population structure was monitored. Badgers in east Offaly are territorial and social groups defend areas of 57-202ha. Each social group contains between 1 and 12 adults (mean = 5.8). In all, 57 setts were identified but no more than 51% were ever in use in the same month. There was one main sett per social group, giving a density of 0.7 km^{-2}, and main setts were about 0.8km apart, indicating that badgers are very common in Offaly. Most setts (55%) were found in hedgerow, which occupies only 3% of the habitat, and their location was little affected by soil type. About four months into the study trapping of badgers began. In all, over the two-year period, 35 adult badgers were caught and three were found dead. By analysis of catch-return curves it is estimated that the total population in 1989 consisted of 51 adults (3.24-5.86 adults km^{-2}). At least 14 social groups used the study area. Boundaries of social group territories were quite resistant to removal of badgers and were still marked and defended even when 51% of the population had been removed. There was evidence, however, of increased transgressions of boundaries at the lower density. By autumn 1990 between 80% and 90% of the original adult population were estimated to have been lost by trapping and natural mortality. The findings of this study are discussed in relation to similar studies.

INTRODUCTION

The Eurasian badger, *Meles meles,* is probably Ireland's most common and widespread carnivore although current information on its distribution and density is incomplete. This situation is clearly improving as the National Badger Survey progresses (Smal 1993). The badger's success has been largely attributed to its ability/willingness to dig setts in a wide range of habitat types (Neal 1986).

We here present information on sett and population densities of an intensively surveyed area in County Offaly. This is of more than local interest since, apart from the detailed information on a particular badger population and the factors influencing sett location, it may also serve to assist in the calibration of the National Badger Survey, which is based primarily on the identification and counting of active setts. Census data on setts have been quite successfully used to produce estimates of the badger population in the United Kingdom (Clements *et al.* 1988; Cresswell *et al.* 1990). Studies such as these evaluated the usefulness of sett densities to predict badger density by referring to more detailed localised studies that mapped territories and estimated badger numbers.

A number of mustelid species, for example otters, may live in groups (Chanin 1985). Such associations, however, are primarily family groupings and are usually short-lived (Powell 1979; Chanin 1985). The Eurasian badger is unique or rare among mustelids, exhibiting gregarious living and associated territorial and social behaviour (Kruuk 1989). The degree of sociality ranges from a close-knit social group or 'clan' system to an almost solitary existence. Some of this variation is attributed to pertaining ecological conditions. It was therefore necessary to establish whether badgers in the area were territorial and, if so, the size and configuration of the individual territories. Average social group size, age structure and sex ratio would provide useful information on how these groups were structured as well as estimates of population density for the area. Establishing the extent of territoriality in badger populations and determining the number and types of setts associated with social groups will aid in explaining census data. Furthermore, determining the number of animals in social groups avoids inappropriate and unreasonable extrapolations.

The study site was situated in east Offaly and was part of an area designated for badger control operations by the Department of Agriculture and Food which were due to start in early 1989. Permission was sought to collect field data on the badgers prior to removal. An attempt was made to estimate the efficiency of the trapping operation and, since animals were removed over a 15-month period, to examine the effect of a substantial reduction in badger numbers on the social structure of the population.

MATERIALS AND METHODS

The study site (approximately 16km^2) is situated in the eastern part of County Offaly, central Ireland, just south of the village of Geashill. Typical of most of the central lowlands, the area is developed chiefly upon gently folded Lower Carboniferous limestone (Herries Davies and Stephens 1978) and consists largely of flat to gently rolling terrain with isolated areas of kame and kettle-hole topography (at altitudes of 70-100m OD) (R. Hammond, unpubl.). The site consists primarily of agricultural land of varying quality and is bounded by a conifer plantation and raised bog to the north-west and mixed woodland and bog to the south. There is a small lake, Raheen Lough (13ha), in the north-east part of the study site. The area experiences a typical Atlantic climate with an annual mean air temperature of 9.3°C. Rainfall occurs throughout the year and there is little snowfall owing to strong westerly winds from the Atlantic (Rohan 1986).

Ten habitat types, based on the criteria outlined by Harris *et al.* (1989), were distinguished in the study area (Table 1). Although there was some crop rotation over the two years the overall abundance of pasture and arable land remained unchanged. Data on the various soil types in the area were acquired from the National Soil Survey of Ireland (R. Hammond, unpubl.).

Table 1. The areas of the ten habitat types recorded in the study area, their relative proportions of the total area and the dominant plant species in each habitat type.

Habitat	Area (ha)	Proportion of total area (%)	Dominant species
Coniferous woodland	64.9	4.13	Norway spruce *Picea abies*; Scots pine *Pinus sylvestris*; larch *Larix decidua*
Broad-leaved woodland	11.15	0.71	Beech *Fagus sylvatica*; oak *Quercus robur*; ash *Fraxinus excelsior*; holly *Ilex aquifolium*
Mixed woodland	26.55	1.69	Beech; oak; ash; pine; rowan *Sorbus aucuparia*; elder *Sambucus nigra*; hazel *Corylus avellana*; birch *Betula pendula*
Scrub	35.66	2.27	Hazel; whitethorn *Crataegus monogyna*; blackthorn *Prunus spinosa*; gorse *Ulex europaeus*; bramble *Rubus* agg.
Raised bog	84.2	5.36	Dominated by *Sphagnum* spp; also bogcotton *Eriophorum angustifolium*; heather *Calluna vulgaris* and Ericaceous spp
Wetground	91.1	5.8	*Juncus* spp; iris *Iris pseudacorus*; meadow buttercup *Ranunculus acris*
Grassland	1049.3	66.8	Varied composition
Arable land	194.8	12.4	Primarily cereals but also root crops and vegetables
Standing water	13.04	0.83	
Total	1570.7		

The activity of every sett was estimated at each visit by counting the number of used and disused entrances. The recent history of activity of each sett was also noted. A sett was classified into one of four categories: main, annexe, subsidiary or outlier, based primarily on its size (no. of entrances) and evidence of use (Thornton 1988; Harris *et al.* 1989). A preliminary survey of the area, carried out by officers of the Department of Agriculture and Food before this study began, had located the majority of the main setts. All other setts were located in the first months of the study. Each sett was assigned to a particular habitat type. For any sett located on the boundary of two habitat types a value of 0.5 setts was assigned to each habitat category.

Two methods were used in the calculation of population size. Assuming that there was little or no movement into the study site over the study period (see discussion), the minimum number present prior to disturbance was estimated by counting numbers caught and animals found dead over the two years (direct enumeration method). By also assuming that badgers remained equally catchable throughout the trapping operation an analysis of catch-effort was used to estimate the original population size. If the trapping operation has a significant effect on population size then by plotting trap-effort (measured as trap-nights) against trapping success (measured as numbers of badgers caught) a curvilinear relationship might be expected, the asymptote of which approximates the total number of catchable badgers present. A number of models were fitted to the catch-effort data (Seber 1982; Krebs 1989). The residual proportion of the population was determined by accumulating trapping returns with natural mortality, using a mortality rate for adults of 25% per annum (Anderson and Trewhella 1985). An estimate of the age of each badger was obtained by a number of different techniques, including cementum annuli incrementation in teeth, the development of the skull and skull size (R. Whelan, pers. comm.). Badgers were reliably classified into only three age categories: cubs, yearlings and animals two years and older.

The coloured bait-marking method of Kruuk (1978) was used to investigate the spatial organisation of the badger population. Territorial activity is at a peak in spring (Kruuk 1978; Cheeseman, Jones *et al.* 1981), with members of contiguous social groups actively demarcating territorial boundaries with latrines and secretions from various scent glands. Badgers in each social group were fed a mixture of peanuts, treacle and small inert plastic pellets (Athlone Extrusions Ltd) on at least ten occasions in the spring of 1989 (March-May) and again in 1990 (Feb.-March). On average, two active setts (range: 1-5), the main sett and usually one subsidiary, were 'baited' per social group. The indigestible pellets could later be detected in the latrines. By using pellets of different colours for the different social groups the boundary latrines of different social groups were identified. More accurate estimates of the territorial boundaries were made by considering the location of marked latrines in relation to badger paths and various landmarks, such as rivers, roads and hedgerows, associated with territorial boundaries. Two types of latrines were distinguished. 'Major' latrines were in use throughout the study and consisted of > 5 faecal droppings or a cumulative total of pits and droppings > 10 at time of inspection. All other defecation sites were termed 'minor' latrines. The similarity of this dichotomy with that of Roper *et al.*'s (1986) 'latrines' and 'temporary defecation sites' allowed the use of major latrines as another parameter in the assessment of territorial boundaries.

The bait-marking method is considered a reliable technique when describing social group territories (Kruuk 1978; Roper *et al.* 1986). The effect of the number of boundary markers on the variance of estimates of territory size was examined to define the minimum number of markers required and the associated variances.

To consider the presence of a social spacing-out mechanism the nearest-neighbour distances (NND) between main setts were determined and their deviation from a random distribution tested. Using the standard formula of Clark and Evans (1954), the mean observed NND (r_a) was compared with the expected NND (r_e) as calculated from the main sett density. Secondly, observed NNDs between the main setts were compared with the distribution of NNDs between a similar number of randomly spaced points (Kruuk 1978). Observed habitat preferences were taken into account and the process repeated three times.

All field data were recorded in notebooks and on 1:10,560 scale maps (Ordnance Survey of Ireland). Areas and distances were measured using a Kontron MOP-VIDEOPLAN• image analysis system and statistical analyses were carried out with the aid of the SAS• package.

RESULTS

Factors influencing density and distribution of setts

Over the two years a total of 57 setts were found in the study area (3.63 setts km^{-2}). Only one new sett was dug although several became inactive (see below). One sett on the periphery of the study area was not regularly checked and so is excluded from the following analyses.

Of the 56 setts checked regularly, 11 (19.6%) were main setts, 4 (7.1%) were annexes, 15 (26.8%) were subsidiary, and 23 (41.1%) were outliers. The other three (5.4%) were derelict and could not be classified (Table 2). Thus, the proportions of the various sett types in the study area were: main setts (1), annexe setts (0.36), subsidiary setts (1.36), outlying setts (2.1), and unclassified setts (0.27). This gives a ratio of main setts to all other sett types of 1 : 4.1. The number of entrances per sett ranged from 1-30 with a maximum of 17 in use at any one time. Main setts had, on average, 11.91 (± 2.8) entrances, annexe setts had 5 (± 4) entrances, subsidiary setts had 3.67 (± 0.31) entrances and outlying setts had 1.41 (± 0.11) entrances (Table 2). The mean numbers of used and disused entrances for each sett category prior to disturbance (1989) were also calculated and are presented in Table 2.

Thirty-five setts were used between spring and autumn 1989 (63.64% of all setts, n = 55 for 1989) including all main setts, giving an overall density of 2.23 active setts km^{-2}. The maximum number of setts active in any month was 28 (51%) and occurred in May 1989. During the year following the first trapping effort the number of active setts declined by 31% for the same period (spring-summer). This represented a significant decrease ($G = 9.95$, $P < 0.005$) and gave a resultant density of 1.53 active setts km^{-2}. The actual change in sett use from 1989 to 1990 was greater (59.5%) as 7 of the 24 setts occupied in 1990 were not in use the previous year. This indicates that, as well as a reduction in the number of setts inhabited, badgers in many cases moved into setts which existed but were unoccupied the previous year. There was an increasing tendency to occupy the same setts in successive years in the order of outlier setts (change = 88.2%) through subsidiary and annexe to main setts (change = 9.1%). Of the 17 setts that were used in both years no variation in activity, as measured by the number of entrances in use in both years, was observed ($t = 0.987$, $P > 0.2$).

Table 2. The number, density, relative abundance and size (± se) of the different sett types in the study area.

	SETT CATEGORY				
	Main	**Annexe**	**Subsidiary**	**Outlier**	**Unclassified**
Number of setts	11	4	15	23	3
Density (km^{-2})	0.65	0.24	0.89	1.37	0.18
Relative abundance*	1	0.36	1.36	2.1	0.27
Number of used entrances	8 ± 1.83	1 ± 0.58	1.75 ± 0.35	0.73 ± 0.11	
Number of disused entrances	3.91 ± 1.68	4 ± 3.51	1.92 ± 0.3	0.68 ± 0.14	
Total number of entrances	11.91 ± 2.8	5 ± 4	3.67 ± 0.31	1.41 ± 0.11	

* Abundance of sett type relative to main setts.

Sett location was influenced by habitat type ($G = 164.3$, $P < 0.001$). No significant relationship was found between the number of setts in a particular habitat and the area of that habitat (Spearman Rank Correlation, $r_s = 0.1605$, $P > 0.05$). No setts were found in raised bog or wet ground, while over half the setts (54.4%) were situated in hedgerows, a habitat category which occupies only 3.2% of the study site (Table 1; Fig. 1). Conversely, only 6 setts (10.5%) were found in grassland, which accounts for over 64% of the study area. Preference or avoidance of specific habitats was determined by constructing confidence intervals for each habitat using the Bonferroni *Z* statistic. Using 80% ($a = 0.2$) simultaneous confidence intervals for all 9 ($k = 9$) habitats available to sett location, 98.9% individual confidence intervals (for each habitat) were constructed (Table 3A). The *Z* value used in constructing these individual confidence intervals was:

$$Z_{(1 - a/2k)} = Z_{(0.989)} = 2.535.$$

Hedgerow appeared to be the only habitat actively selected. Both grassland and arable were clearly avoided (Table 3). A similar avoidance of raised bog and wet ground was inferred. Although a relatively high proportion of setts were in coniferous woodland, selection was not proven. The results for broad-leaved and mixed woodland and for scrub were more ambiguous. Although these three habitats were not avoided it was not possible to determine whether they were actively selected. As the lower confidence limit for each habitat tended to zero, owing to the low availability of the habitat (in all cases 3%), only avoidance could have been detected.

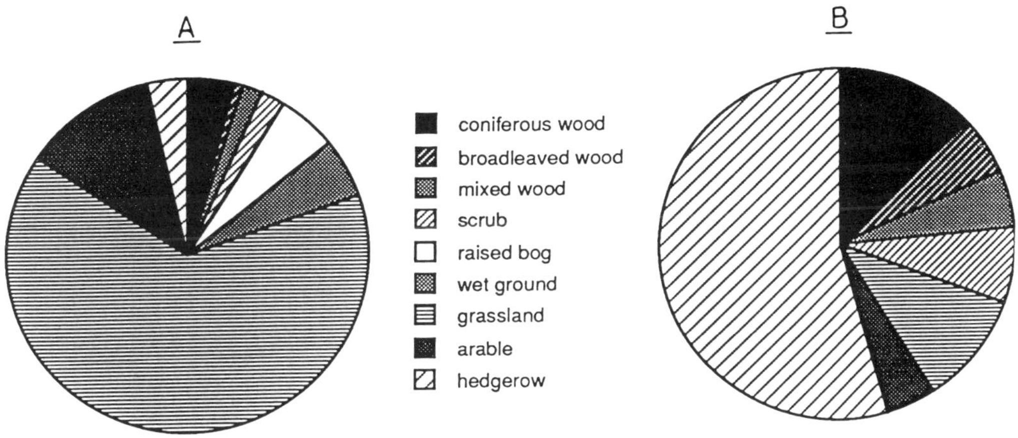

Fig. 1. Habitats and location of setts. (A) Habitat types (excluding standing water) in the study area (15.72km^2). (B) Subdivision of setts (n – 57) by habitat.

All three woodland categories were combined to see if woodland in general was actively selected for: 80% simultaneous and 98.6% individual ($Z = 2.45$) confidence intervals were constructed for the resulting 7 (k) habitat categories. Badgers were indeed actively selecting woodland, with over 23% of all setts in woodland of one type or other (Table 3B). This preference for hedgerows and woodland suggests an active selection for cover when choosing a site in which to dig a sett.

Over 85 % of all setts were located in habitats offering a moderate to high degree of cover all year round (woodland, hedgerow and scrub), with nearly 15% in open habitat (pasture and arable), further indicating a significant selection for cover in location of sett ($G_{\text{including bog and wet ground}} = 149.8$, $P << 0.001$; $G_{\text{excluding bog and wet ground}} = 139$, $P << 0.001$). The majority of setts in open habitats had their origins in hedgerows, scrub or woodland, but some entrances are now in open fields where hedgerows or other cover have been removed during reclamation. One sett was in a cave at the crest of a kame in pasture, while another was situated 18m into a large arable field, withstanding the attentions of the plough on several occasions.

About 16% (9/56) of all setts were at the interfaces of two habitat types. In all cases the habitat combination offered both cover (scrub, hedgerow or coniferous woodland) and open pasture. Such an observation is not sufficient to indicate whether badgers select habitat interfaces for digging setts as the position of such setts may be an artefact of changing land use (see above). To investigate this phenomenon all setts located in uniform habitat blocks with a minimum width of 20m were investigated to see if they were situated towards the habitat edge (≤ 5m from next habitat). This excluded setts sited partly or completely in hedgerows, which by their very nature are ≤ 5m from the neighbouring habitat. Of the remaining 24 setts, 15 (62.5%) were ≤ 5m from the adjoining habitat and only 9 (37.5%) were > 5m from the adjoining habitat. Most (90%) of the setts that were close to another habitat were in habitat offering good all-year-round cover (woodland and scrub) abutting pasture.

There was no significant selection for particular soil types in the choice of site for sett location ($G = 5.287$, $P > 0.05$); the number of setts in each soil type was highly correlated with the area of the various soil series ($r_s = 0.8512$, $P < 0.01$). When sett location was analysed with respect to soil drainage class, although no setts were found in the most poorly drained soils, no significant preferences for particular soils were observed ($X^2 = 1.8645$, $P > 0.5$).

Table 3. Habitat selection in the location of setts: preference for or avoidance of individual habitats: (A) all habitats; (B) the 7 habitats when all woodland categories are combined.

A

Habitat	Proportion of area and hence expected proportion of setts	Proportion of setts actually observed	Upper and lower 80% confidence limits	
Coniferous woodland	0.0416	0.1316	0.0181	0.2451
Broad-leaved woodland	0.0072	0.0526	0	0.1276
Mixed woodland	0.017	0.0526	0	0.1276
Scrub	0.0229	0.0702	0	0.156
• Raised bog	0.0541	0	-	-
• Wet ground	0.0563	0	-	-
* Grassland	0.6486	0.1053	0.0023	0.2803
* Arable	0.1205	0.0439	0	0.1127
• Hedgerow	0.0319	0.5439	0.3767	0.7111

B

Habitat	Proportion of area and hence expected proportion of setts	Proportion of setts actually observed	Upper and lower 80% confidence limits	
• Woodland	0.0658	0.2368	0.0988	0.3748
Scrub	0.0229	0.0702	0	0.1532
• Raised bog	0.0541	0	-	-
• Wet ground	0.0563	0	-	-
* Grassland	0.6486	0.1053	0.0057	0.2049
* Arable	0.1205	0.0439	0	0.1104
• Hedgerow	0.0319	0.5439	0.3823	0.7055

* Habitats actively avoided. • Habitats actively selected. • Habitats that were clearly avoided but avoidance not proven.

When sett location was analysed with respect to surrounding topography within an area that might include the home range of the occupants, no significant preference for particular terrain types was found ($G = 0.476$, $P > 0.05$). Nevertheless, 80.7% (46/57) of setts were dug into local slopes. These include hedgerow and dyke banks and railway embankments as well as generally sloping topography. Of all setts dug in man-made slopes, 90% (38.5/43) were in areas of moderate to good cover, and 75% (29/38.5) of these were in linear areas of cover such as hedgerows.

The mean nearest-neighbour distance (r_a) between the 11 main setts in the study area was 0.801km. This was compared to the expected mean nearest-neighbour distance (r_e = 0.60km), calculated from the main sett density. Main setts were, on average, 1.34 times as far apart as expected under conditions of randomness and this difference was significant ($C = 2.155$, $P < 0.01$). Secondly, observed nearest-neighbour distances between the 11 main setts were compared with the distribution of nearest-neighbour distances between 11 randomly spaced points. Observed habitat preferences were taken into account by plotting 4 points in cover, 6 in agricultural land (excluding wet ground) and one on a boundary of cover and agricultural land. Main setts were further apart (minimum NND = 526m; $\bar{x}$= 801m) than would be expected if they were randomly distributed according to habitat preference (minimum NND = 104.2m; $\bar{x}$ = 556.7m).

Social organisation

The study site was occupied by at least 14 social groups, each occupying and actively demarcating a territory. Complete boundaries were described for the 8 'core' groups only (Fig. 2). The mean territory size of these 8 groups was 87.4ha (se = 8.32, range: 57-118ha, Table 4). The mean greatest linear distance across territories (D_{max}) was 1.39km (se = 0.1) and ranged from 1.07-1.93km for the 8 core groups.

A minimum of 8 regularly spaced points along the perimeter sufficiently explains territory shape, because even when only 8 fixes of a possible 15 were used to define the boundary of the irregularly shaped territory of the 'Fox' social group all combinations of 8 fixes yielded a coefficient of variance of only 19%. As the number of latrines used to define the territories of the 8 core groups ranged from 9 to 26, apart from other criteria, territories were adequately defined by the above techniques. Using data from these core groups a badger social group had, on average, 4.375 setts (range: 1-8) comprising 1 main sett, 0.25 annexe setts, 1.375 subsidiary setts, 1.625 outlying setts and 0.125 derelict setts. During peak activity (May 1989) the number of active setts per social group ranged from 1 to 5 ($\bar{x}$ = 2.38).

By the spring of 1990 the social organisation of the population had changed. The complete territorial boundaries of six contiguous social groups were identified. Five of these existed the previous year, although most had changed. The other was mainly the product of the combination of two social groups from 1989 (Fig. 3). The mean territory size was 102ha (se = 19.8) with a range of 57-202ha (Table 4). The mean greatest linear distance across a territory was 1.53km (se = 0.12) and ranged from 1.3km to 2.04km. Overall there was no significant change in territory size between 1989 and 1990 (Mann-Whitney, $U = 27.5$, $P > 0.2$) although some territories increased and others decreased, in some cases substantially, while one remained the same (Table 5).

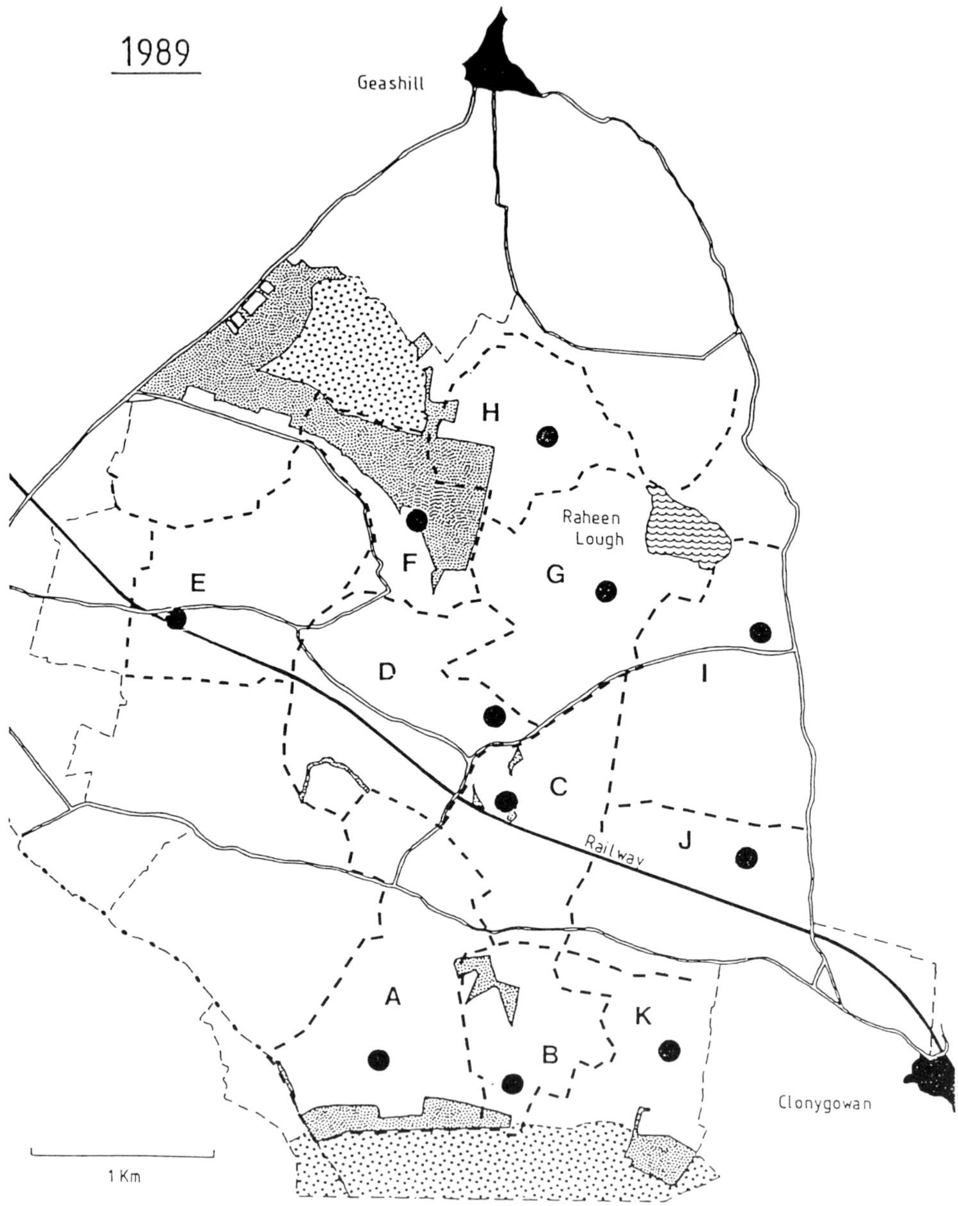

Fig. 2. Territorial boundaries and main setts of the social groups in the study site in 1989. Eleven main setts (•) and the boundaries (——) of at least 14 social groups were identified. Pecked lines (----) and roads (═) delineate the study area; stippled areas = raised bog; grey areas = woodland; dot-dash line = county border.

Table 4. Size, perimeter length, greatest linear distance and sett composition of the territories of the eight central badger groups in 1989 (A) and the six central groups in 1990 (B).

A

Social group	Area	Dmax	Perimeter length	Setts					
	(ha)	(km)	(km)	Main	Annexe	Subsidiary	Outlier	Unclassified	Total
A Fox	117.6	1.93	5.92	1	1	2	1	-	5
B Otter	57.01	1.07	3.59	1	-	-	-	-	1
C Harvey's	86.74	1.56	4.35	1	-	1	1	-	3
D Mill Race	107.2	1.31	5.17	1	-	2	2	-	5
E Railway	90.51	1.36	4.39	1	-	2	5	-	8
F Plantation	57.52	1.41	4.08	1	1	2	2	1	7
G Lough	110.1	1.16	5.44	1	-	-	-	-	1
H Red Hill Town	72.34	1.29	3.94	1	-	2	2	-	5
Mean	87.38	1.39	4.61	1	0.25	1.375	1.625	0.125	4.375
± sd	23.52	0.27	0.81						
(± se)	(8.2)	(0.1)	(0.29)						

B

Social group	Area	Dmax	Perimeter length	Setts					
	(ha)	(km)	(km)	Main	Annexe	Subsidiary	Outlier	Unclassified	Total
A-B Fox-Otter	201.6	2.04	7.3	2	1	2	1	-	6
C Harvey's	94.9	1.71	5.55	1	-	1	1	-	3
D Mill Race	72.5	1.3	3.63	1	-	2	3	-	6
E Railway	90.86	1.37	4.54	1	-	2	5	-	8
F Plantation	57.5	1.41	4.10	1	1	2	2	1	7
H Red Hill Town	94.4	1.33	4.27	1	-	2	2	-	5
Mean	102	1.53	4.9	1.167	0.333	1.833	2.333	0.167	5.833
± sd	48.5	0.29	1.35						
(± se)	(19.8)	(0.12)	(0.58)						

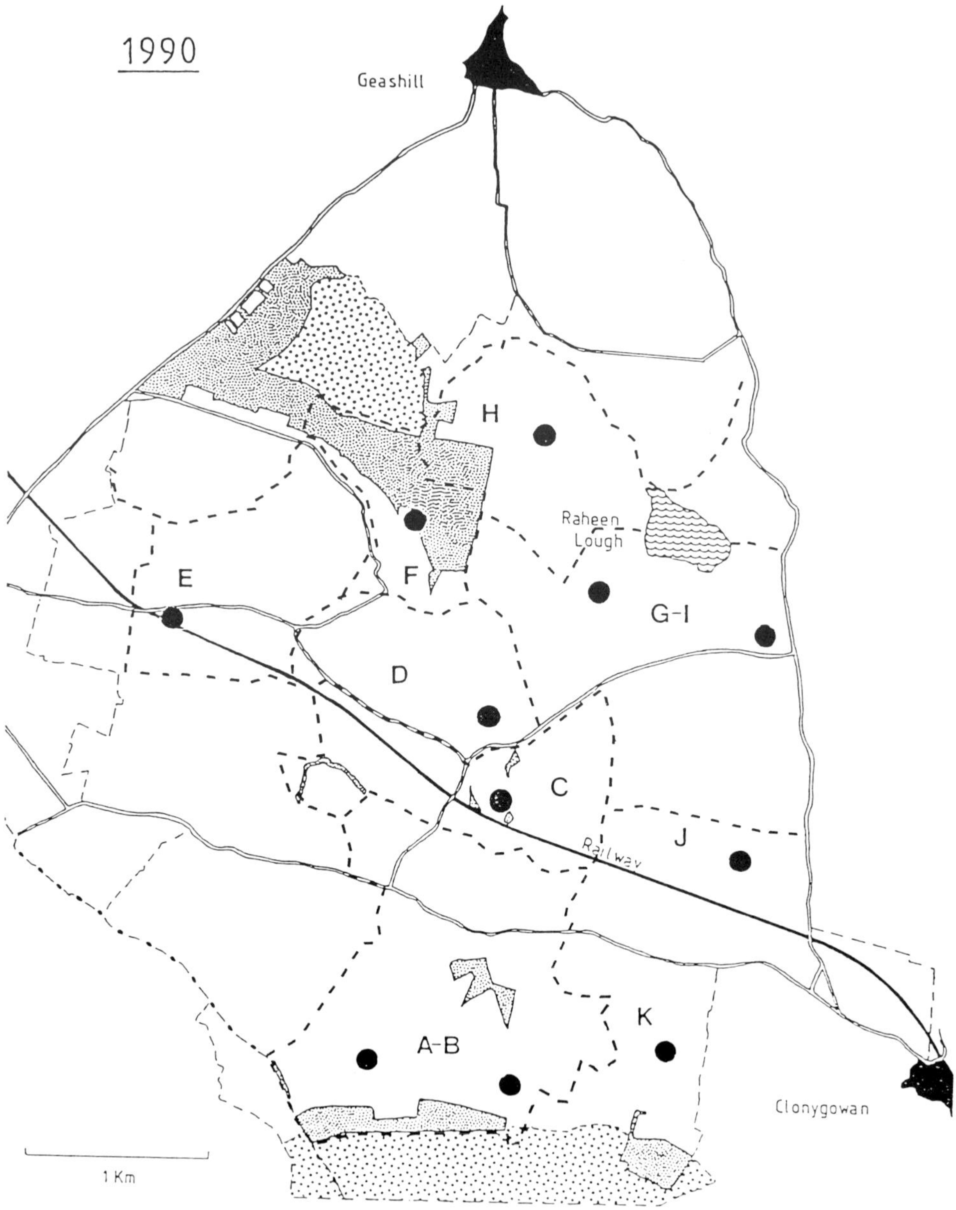

Fig. 3. Territorial boundaries and main setts of the social groups in the study site in 1990. Key as in Fig. 2.

Table 5. Change in territory size and configuration of the five social groups which remained intact during the study period (i.e. to spring 1990).

Social group	Area 1989 (ha)	Area 1990 (ha)	Area lost (ha)	Area gained (ha)	Total area occupied over 2 years	Change from 1989 - 1990 (%)	Territory size increase/decrease (ha)	(%)
Harvey's	86.74	94.9	30.87	39.1	125.84	55.6	+ 8.16	9.41
Mill Race	107.2	72.51	45.88	11.19	118.39	48.21	- 34.69	32.36
Railway	90.51	90.86	0	0.35	90.86	0.39	+ 0.35	0.37
Plantation	57.52	57.52	0	0	57.52	0	0	0
Red Hill Town	72.34	94.37	0	22.03	94.37	23.34	+ 22.03	30.45

Population estimates

An estimate of minimum population density in the study area is based on the number of badgers known to be alive prior to disturbance by using the number of animals trapped or found dead for those social groups whose territories were completely defined in both years. Assuming little or no movement in or out of this area ($7km^2$), a minimum of 31 badgers, 28 of which were adults, occupied the core area in 1989, yielding a density of 4.4 badgers km^{-2} and 4 adults km^{-2} respectively. Five of the 8 core groups remained intact over the two-year study, despite the change in boundary configuration in this time (Fig. 3). Estimates of social group size are based on these 5 groups over the 2 years. The number of adults per social group ranged from 1 to 8 with a mean of 4. There was no correlation between social group size and territory size ($r = -0.48$, $P > 0.05$, and $r_s = 0.375$, $P > 0.05$; Fig. 4). As the number of cubs recorded was unusually low ($n = 4$), only sex ratios of adults were calculated. The sex ratio for the 8 central groups, as calculated from trapping data collected over the two years, was 1 : 1.45 (11 males : 16 females); this difference was not significant ($X^2 = 0.9259$, $P > 0.05$). Likewise, although there were more females in the study area as a whole (1 : 1.54, 13 males : 20 females), the sex ratio did not differ significantly from unity ($X^2 = 1.485$, $P > 0.05$). Despite overall parity, the sex ratio varied between social groups, for example 1 male and five females were recorded from Harvey's whereas 3 males and 1 female were recorded from the Railway group.

Assuming the population of badgers is finite, based on the curvilinear relationship between trapping effort and success (Fig. 5a), it is possible to employ catch-effort methods in estimating initial population size (N). The regression of catch per unit effort against cumulative catch (Leslie plot) for each week from the combined trapping data of 1989 and 1990 yielded an estimate of N = 48 ($P < 0.001$, $r^2 = 0.725$; Fig 5b). This was taken as a first approximation of N as the variance about the regression line was substantial. By combining trapping data from 1989 and 1990 (spring) the reciprocal plot of trap-effort against trapping success gave a stronger relationship ($P < 0.001$, $r^2 = 0.935$) and estimated a B_{max} (N) of 51 (Fig. 5c). This value represents an estimate of the total number of badgers, or more correctly, as 97% of the badgers caught in this period were adults (34/35), the total number of adult badgers that could be caught in the area.

In 1989 26 adults were caught and one was found dead. This represents a 51% trapping success and a 53% total recovery in the first year of all adult badgers known to be present. By the summer of 1990 34 adults were caught; however, only 33 were adults in the spring of 1989, and 2 were found dead. This represents a 64.7% catch success and a 68.6% total recovery of the adults present in 1989. Three adults were caught and one found dead during autumn 1990. Three of these 4 animals had reached adulthood prior to disturbance. Thus, a total of 35 adults were caught and 3 recovered dead since trapping began, representing a 68.6% catch success and a 74.5% total recovery respectively.

Assuming a mean mortality rate for adults (> 1 year) of 25% per annum, a minimum of 12 adult badgers may have died in the field in the interval (15 months) since trapping began. As 3 adults were found dead, a further 9 may have been unrecorded. Therefore 47 adult badgers may be accounted for. This represents 92% of the adults present in the study area before disturbance. However, a value between 80% and 90% is a more realistic estimate as some of the adults destined to die during the study would have been caught, and immigration to the population in the latter months cannot be discounted.

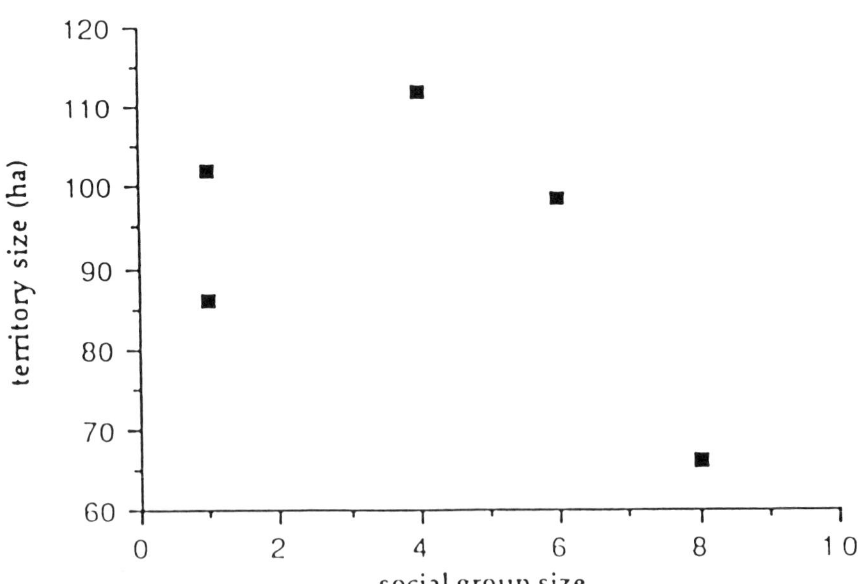

Fig. 4. The relationship between territory size and social group size (minimum estimates): $r = -0.61$, $P > 0.05$.

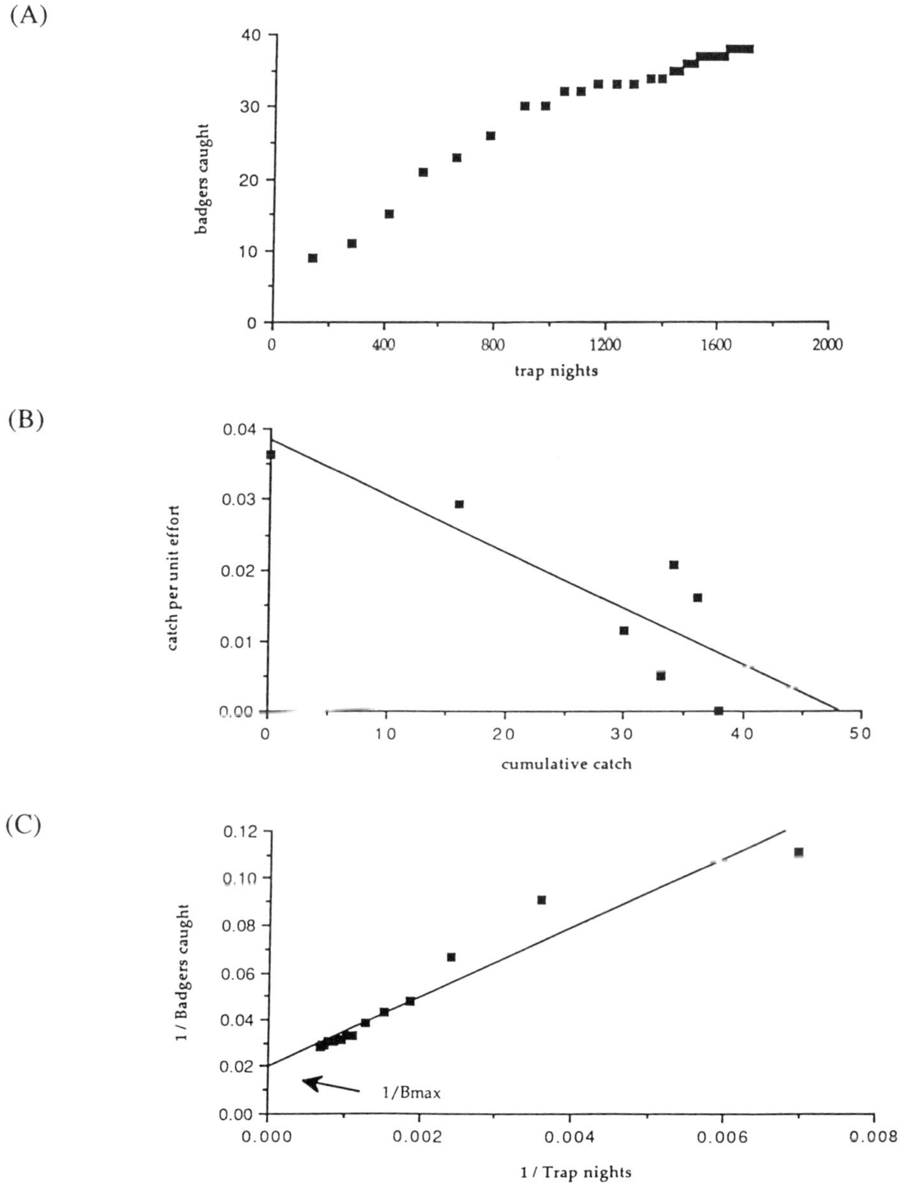

Fig. 5. Estimation of population size. (A) Cumulative catch of badgers in relation to trapping effort (trap nights). (B) Trapping success (badgers caught per week) in relation to cumulative catch (Leslie plot), $r = 0.851$, $P < 0.001$. The intercept on the ordinate gives an estimate of the total population, $N = 48$. (C) Double-reciprocal of data in (A), $r = 0.967$, $P < 0.001$. The inverse of the intercept on the abscissa is an estimate of the total population, $B_{max} = 51$.

If the original population contained 51 adults, then adult density in the study area (15.72km^2) was 3.24 km^{-2}. Based on a total recovery at this stage of 68.6%, 41 adult badgers were present in the area occupied by the 8 central groups. This represents a local density of 5.86 adult badgers km^{-2}. Twenty animals which were adults in the spring of 1989 were recovered from the five social groups which remained intact over the two-year period. Assuming a 68.6% total recovery in this time, some 29 adults would have been present in these five contiguous groups prior to disturbance, giving a mean number of adults per social group of 5.8 with a range of 1-12.

Badger movements within the population

The badger population may be classified as of medium density. The spatial organisation into distinct social groups was well defined, as evidenced by conspicuous territorial demarcation. Linear distance measurements from setts where bait was laid (usually main setts) to latrines where the bait was found revealed that badgers moved up to 2.14km away from the sett in one night. The coloured pellets were found between 1m and 1000m (1km) from baited setts in 1989 ($\bar{x}$ = 413m). In 1990 the pellets were found up to 2137m (2.14km) from sett of origin with a mean of 610m. On average, pellet returns in 1990 were significantly further away from point of origin than in 1989 ($t = 2.407$, $P < 0.05$). Coloured pellets were sometimes found inside the territories of adjacent social groups, indicating that badgers may transgress territorial boundaries. In some cases pellets of two different colours were found in the same faecal dropping, indicating that that badgers had visited two main setts in the same night. Linear distance estimates indicated that badgers feeding at two setts successively and subsequently defecating in a latrine moved a minimum of 1.36km-3.13km. There was a significantly greater number of recorded territorial transgressions (4 compared to 1) and 'double' colour returns (3 compared to 1) in 1990 compared to 1989 ($G = 4.93$, $P < 0.05$).

DISCUSSION

Eleven main setts were found in the study area (15.72km^2), all of which were active in 1989. This is equivalent to a main sett density of 70/100km^2, thus ranking badger setts as 'very common' for the east Offaly region according to the criteria of the National Sett Survey in Britain (Neal 1986; Clements *et al.* 1988). The average size and the mean number of used and disused entrances of the four sett categories are within the range reported for Britain and indeed are very close to the national average (Cresswell *et al.* 1990).

It is not known if there exists a hierarchy of decisions by which badgers respond to ecological factors and biological influences when choosing where to excavate a sett. Badgers showed particular habitat preferences when digging a sett. The high proportion of setts found in hedgerow, woodland and scrub (85%) indicates a selection for, among other factors, cover. Furthermore, of the four setts situated completely in the open, one is in a cave and all acquire cover on a seasonal basis. These findings confirm the general observation that badgers select secluded sites in order to conceal those activities carried out close to the sett (Neal 1972; Kruuk 1978; Cheeseman, Jones *et al.* 1981). In contrast to these and many other studies (e.g. Skinner *et al.* 1991), however, hedgerows replaced deciduous woodland as the most important habitat for sett location. The unusually high proportion of setts in hedgerow is due, in part, to the lack of deciduous woodland (0.72%) and the relative abundance of hedgerow. The areas of fields are smaller and the hedgerows larger (average width = 2.5m) relative to those in Britain (average width = 1.5m; Cresswell *et al.* 1990). The proportion of setts located in open situations (15%) may seem quite high (Neal 1972), but the majority of these are only partly in the open and this primarily by default, as several were left exposed owing to recent hedgerow removal. The fact that most of these setts, including those completely in the open, were occupied indicated that badgers were reluctant to vacate setts despite the lack of continuous cover. Although setts were rarely far from the next habitat, setts in sizeable tracts of uniform habitat tended to be situated towards the habitat edge. The majority are in cover near pasture. Such preferences suggest selection for factors such as plentiful and varied food supply and seclusion as well as cover (Neal 1972).

Badgers generally avoid coniferous woodland when choosing a site to locate a sett. Such habitats lack ground cover and adequate food and bedding (Neal 1972; 1986; Kruuk and Parish 1987). But in Offaly over 13% of setts were located in coniferous woodland, a habitat type which accounted for only 4.2% of the study area. This preference may be explained by the greater seclusion from human activity found in the coniferous plantation and the structural support offered to setts by the well-established root system of the stands of mature pine, *Pinus sylvestris,* and spruce, *Picea abies*. The possibility that some setts were established prior to the creation of the coniferous woodland can be discounted primarily because these plantations were developed chiefly upon raised bog, a habitat actively avoided by badgers. A similar selection for coniferous woodland has been observed in Scotland (Cresswell *et al.* 1989).

In certain areas the properties of soils may be a primary determinant of sett location. Two of the most important are ease of excavation ('diggability' *sensu* Thornton 1988) and moisture content (Likhachev 1956; Dunwell and Killingley 1969; Neal 1972; 1986; Kruuk 1978; Thornton 1988). Eight soil types occur in the study site but no overall selection for particular soils was detected. However, badgers avoided the two wettest habitats. No setts were found in either raised bog or wet ground, which together represented 11% of the area available to sett location. This indicates that the level of the water table is an important ecological factor in the choice of site for a sett. Of 635 setts located in the whole of east Offaly (autumn 1990) none were in bog (Eves 1993). The high proportion of setts found in slopes (80.7%) further indicates a preference for a well-drained substratum (Neal 1972; Clements *et al.* 1988). Digging a sett in a slope also facilitates the removal of excavated soil and access to favourable strata (Neal 1972). In the majority of cases slopes were man-made, where the soils are less consolidated.

Although the density of main setts was similar to the upper values from Britain, total sett density (3.63 km^{-2}) was much higher than the corresponding British estimates (Cresswell *et al.* 1990). The relative abundance of secondary setts (annexe, subsidiary and outlier) to main setts (1 : 4.1) was far greater than comparable areas (land classes) in Britain ($\bar{x}$ = 1 : 2.12; Cresswell *et al.* 1990). This appears to be related to the distribution as well as the availability of cover. The overall abundance of cover is the same or even greater in many parts of Britain (Neal 1986; Clements *et al.* 1988; Cresswell *et al.* 1990) compared to Offaly. However, owing to the large size of fields, successive units of cover (hedgerow and woodland) are further apart. Furthermore, taking into consideration that the location of a sett within a territory is probably influenced by its proximity to existing setts, the number of opportunities available to dig a sett in parts of Britain may be more limited. In east Offaly by contrast, despite a lower availability of cover, the patchwork of small fields bounded by relatively wide hedgerows may create a situation where all the requirements for sett location, intrinsic social factors as well as ecological parameters, are met more often. It may be hypothesised that maintaining a number of setts beyond the immediate requirements of the social group may be of adaptive significance under conditions of disturbance and/or destruction of setts. Examination of reproductive output of social groups in relation to *inter alia* the spectrum of setts they possess could test this hypothesis.

The distribution of setts is also influenced by the presence of other badgers. Badgers in the study site were territorial and this accounts for the non-random spacing of main setts, even when habitat preferences were taken into account. In central Ireland badgers are territorial. The population is organised into distinct social groups which conspicuously demarcate territorial boundaries. The mean territory size prior to disturbance of 87.4ha was much greater than values recorded in the high-density areas of Gloucestershire, south-west England ($\bar{x}$ = 22-25ha), by Cheeseman, Jones *et al.* (1981), similar to sizes found in Cornwall ($\bar{x}$ = 74.7ha; *ibid.*), Avon ($\bar{x}$ = 73.7ha; *ibid.*) and Staffordshire ($\bar{x}$ = 104 ha; Cheeseman, Little *et al.* 1985), but considerably smaller than average territory sizes of 171ha in the low-density areas of Scotland, where territories may exceed 300ha (Kruuk and Parish 1982; 1987). The density of adult badgers in Offaly, based on catch-effort estimates, of 3.24-5.86 km^{-2} is considerably lower than that in the Cotswold escarpment region of Gloucestershire (19.7 km^{-2}) based on total catch estimates (Cheeseman, Jones *et al.* 1981), but markedly higher than the low densities found in parts of Scotland (2.2 km^{-2}; Kruuk and Parish 1987), Essex (0.54 km^{-2}; Skinner *et al.* 1991), the Netherlands (1 km^{-2}; Wijngaarden and Peppel 1964) and France (1.6 km^{-2} ; Mouches 1982). Although methods used to estimate density vary among studies, the density of adults in Offaly is similar to medium-density populations in Staffordshire (6.2 km^{-2}; Cheeseman, Little *et al.* 1985) and the urban population in Bristol (4.4-7.5 km^{-2}; Harris and Cresswell 1987). The mean number of adult badgers per social group (5.8) is similar to values recorded in several studies in Britain where the social group size for a number of low- to high-density areas averaged 5.03 adults/social group (Cheeseman, Jones *et al.* 1981; Cheeseman, Little *et al.* 1985; Cheeseman, Wilesmith *et al.* 1987; Kruuk and Parish 1982; 1987; Harris and Cresswell 1987; Hofer 1988). Assuming an equal sex ratio at birth (Anderson and Trewhella 1985) the preponderance, albeit non-significant, of females in the adult population may indicate sex-specific differences in mortality. The rigours of dispersal may expose adult boars to increased risk of mortality on roads or from intra-specific aggression (Cheeseman, Wilesmith *et al.* 1987; Harris and Cresswell 1987). Furthermore, a number of studies have recorded higher prevalences of *Mycobacterium bovis* infection and associated mortality in males compared to females (Gallagher and Nelson 1979; Cheeseman, Wilesmith *et al.* 1988; 1989). Adult males appear to be more actively involved in territorial defence and may therefore be at greater risk of contracting infection through bite wounding (Cheeseman, Wilesmith *et al.* 1988).

Despite a dramatic reduction in badger numbers owing to trapping, the spatial organisation of the population was retained during this period and, although there was considerable variation in boundary configuration between years, territory size did not change significantly. Such findings support the view that badgers are 'contractors', and maintain the smallest economically defensible area which encompasses sufficient resources for year-round survival and reproduction of a single animal or breeding pair (Kruuk and Macdonald 1985). If resources such as food are patchily distributed, both temporally and spatially, this allows the development of larger groups. The reduction in numbers in the study area was not sufficient to abolish territorial boundaries, as enough animals remained (> 1 per social group) to maintain them. The lack of a relationship between group size and territory size may indicate a patchy distribution of limiting resources according to the 'contractor' theory (Kruuk and Parish 1982).

Although movements across territorial boundaries were detected in both years, there were significantly more boundary transgressions and colour markers were significantly further away from setts of origin in 1990 than in 1989. Thus badgers moved greater distances away from setts despite the lack of a change in territory size. As there was no significant change in land use over the two years, the increased movement between territories after the first trapping operation cannot be entirely attributed to food shortage. On the contrary, food might now have been more abundant. Thus disturbance disrupted the organisation of the various social groups. The increased movement may indicate opportunistic transgressions across less regularly defended boundaries, or the extension of male ranges to encompass two contiguous territories (Kruuk 1978; Cheeseman and Mallinson 1980). Despite the greater distances travelled post-trapping, the number of recorded territorial transgressions were relatively low and on only one occasion were coloured pellets found in a territory not adjacent to the territory of origin. Furthermore, the extensive tracts of bog to the north and south of the study site (Fig. 2) constitute effective barriers to badger movement, and emigration from or immigration to the study site over the two years was considered to have been limited. Finally, genetic findings suggested that gene flow between groups in the population was limited (O'Corry-Crowe 1992).

Estimates of trapping success indicate that 71% of the adult badgers in the study area prior to disturbance (spring 1989) were caught over the 15-month period. When natural mortality is taken into account, over 90% of the adults were lost to the population. However, since the trapping effort was episodic (3 sessions over 15 months) it is possible that cubs not trapped in the earlier sessions survived to be recruited into the adult population. Also, as time passes, immigration of badgers into the void left by trapping becomes increasingly likely (Cheeseman, Cresswell *et al.* 1988). Since in high-density populations it may take up to 10 years to re-establish pre-trapping density (Cheeseman, Mallinson *et al.* 1993) it is predicted that if no further trapping operations are carried out in this area and the surrounding region, recovery of the badger population to former numbers may take at least 5 years.

ACKNOWLEDGMENTS

We would like to thank: Declan Bray, Michael Hennesy, Jim Lyons, and the staff of the Department of Agriculture and Food, Tullamore, for their assistance in this research; members of the Department of Zoology, UCD, for their help in the field; and all landowners and farmers in the area, without whose continued cooperation this study would not have been possible. Thanks also to the Brennan family, Clonygowan, who provided accommodation for one of us (GO'CC) during fieldwork. We are grateful to Dr R. Hammond (Teagasc) for allowing us to use unpublished soil data for the area. The study was partly funded by the Department of Agriculture and Food.

REFERENCES

ANDERSON, R.M. and TREWHELLA, W. 1985 Population dynamics of the badger (*Meles meles*) and the epidemiology of bovine tuberculosis (*Mycobacterium bovis*). *Philosophical Transactions of the Royal Society, London*, Series B, **310**, 327-81.

CHANIN, P.R.F. 1985 *The natural history of otters*. Beckenham. Croom Helm.

CHEESEMAN, C.L. and MALLINSON, P.J. 1980 Radio tracking in the study of bovine tuberculosis in badgers. In C.I. Amlaner and D.W. Macdonald (eds), *A handbook of biotelemetry and radio tracking*, 646-56. Oxford. Pergamon Press.

CHEESEMAN, C.L., CRESSWELL, W.J., HARRIS, S. and MALLINSON, P.J. 1988 Comparison of dispersal and other movements in two badger (*Meles meles*) populations. *Mammal Review* **18**, 51-9.

CHEESEMAN, C.L., JONES, G.W., GALLAGHER, J. and MALLINSON, P.J. 1981 The population structure, density and prevalence of tuberculosis (*Mycobacterium bovis*) in badgers (*Meles meles*) from four areas in south-west England. *Journal of Applied Ecology* **18**, 795-804.

CHEESEMAN, C.L., LITTLE, T.W.A., MALLINSON, P.J., PAGE, R.J.C., WILESMITH, J.W. and PRITCHARD, D.G. 1985 Population ecology and prevalence of tuberculosis in badgers in an area of Staffordshire. *Mammal Review* **15**, 125-35.

CHEESEMAN, C.L., MALLINSON, P.J., RYAN, J. and WILESMITH, J.W. 1993 Recolonisation by badgers in Gloucestershire. In T.J. Hayden (ed.), *The badger*, 78-93. Dublin. Royal Irish Academy.

CHEESEMAN, C.L., WILESMITH, J.W., RYAN, J. and MALLINSON, P.J. 1987 Badger population dynamics in a high density area. *Symposium of the Zoological Society of London* **58**, 279-94.

CHEESEMAN, C.L., WILESMITH, J.W., STUART, F.A. and MALLINSON, P.J. 1988 Dynamics of tuberculosis in a naturally infected badger population. *Mammal Review* **18**, 61-72.

CHEESEMAN, C.L., WILESMITH, J.W. and STUART, F.A. 1989 Tuberculosis: the disease and its epidemiology in the badger, a review. *Epidemiology and Infection* **103**, 113-25.

CLARK, P.J. and EVANS, F.C. 1954 Distance to nearest neighbour as a measure of spatial relationships in populations. *Ecology* **35**, 445-53.

CLEMENTS, E.D., NEAL, E.G. and YALDEN, D.W. 1988 The National Badger Sett Survey. *Mammal Review* **18**, 1-9.

CRESSWELL, P., HARRIS, S., BUNCE, R.G.H. and JEFFERIES, D.J. 1989 The badger (*Meles meles*) in Britain: present status and future population changes. *Biological Journal of the Linnean Society* **38**, 91-101.

CRESSWELL, P., HARRIS, S. and JEFFERIES, D.J. 1990 *The history, distribution, status and habitat requirements of the badger in Britain.* Peterborough. Nature Conservancy Council.

DUNWELL, M.R. and KILLINGLEY, A. 1969 The distribution of badger setts in relation to the geology of the Chilterns. *Journal of Zoology, London* **158**, 204-8.

EVES, J. 1993 The East Offaly Badger Research Project: an interim report. In T.J. Hayden (ed.), *The badger,* 166-73. Dublin. Royal Irish Academy.

GALLAGHER, J. and NELSON, J. 1979 Causes of ill health and natural death in badgers in Gloucestershire. *Veterinary Record* **105,** 546-51.

HARRIS, S. and CRESSWELL, W.J. 1987 Dynamics of a suburban badger (*Meles meles*) population. *Symposium of the Zoological Society of London* **58**, 295-311.

HARRIS, S., CRESSWELL, P. and JEFFERIES, D.J. 1989 *Surveying badgers.* London. The Mammal Society.

HERRIES DAVIES, G.L. and STEPHENS, N. 1978 *Ireland.* London. Methuen.

HOFER, H. 1988 Variation in resource presence, utilisation and reproductive success within a population of European badgers (*Meles meles*). *Mammal Review* **18**, 25-36.

KREBS, C.J. 1989 *Ecological methodology.* New York. Harper and Row.

KRUUK, H. 1978 Spatial organisation and territorial behaviour of the European badger (*Meles meles*). *Journal of Zoology, London* **184**, 1-19.

KRUUK, H. 1989 *The social badger: ecology and behaviour of a group-living carnivore* (*Meles meles*). Oxford University Press.

KRUUK, H. and MACDONALD, D. 1985 Group territories of carnivores: empires or enclaves. In R.M. Sibly and R.H. Smith (eds), *Behavioural ecology: ecological consequences of adaptive behaviour,* 52-36. Oxford. Blackwell Scientific Publications.

KRUUK, H. and PARISH, T. 1982 Factors affecting population density, group size and territory size of the European badger (*Meles meles*). *Journal of Zoology, London* **196**, 31-9.

KRUUK, H. and PARISH, T. 1987 Changes in the size of groups and ranges of the European badger (*Meles meles* L.) in an area of Scotland. *Journal of Animal Ecology* **56**, 351-64.

LIKHACHEV, G.N. 1956 Some ecological traits of the badger of the Tula Abatis broadleaf forests. In P.B. Jurgensen (ed.), *Studies on mammals in government Preserves.* Moscow. Ministry of Agriculture, USSR. Translation by Israel Program for Scientific Translations, Jerusalem.

MOUCHES, A. 1982 Le Blaireau Européen (*Meles meles*): biologie et eco-éthologie. *Bulletin Mensuel de l'Organisation Nationale de la Chasse,* 21-8.

NEAL, E.G. 1972 The National Badger Survey. *Mammal Review* **2**, 55-64.

NEAL, E.G. 1986 *The natural history of badgers.* Beckenham. Croom Helm.

O'CORRY-CROWE, G. 1992 The social and genetic structure of a badger (*Meles meles* L.) population in central Ireland. Ph.D. thesis (unpublished), National University of Ireland.

ROHAN, P. K. 1986 *The climate of Ireland.* Dublin. Stationery Office.

ROPER, T.J., SHEPARDSON, D. J. and DAVIES, J. M. 1986 Scent marking with faeces and anal secretion in the European badger (*Meles meles*): seasonal and spatial characteristics of latrine use in relation to territoriality. *Behaviour* **97**, 94-117.

SMAL, C. 1993 The National Badger Survey: preliminary results for the Irish Republic. In T.J. Hayden (ed.), *The badger,* 9-22. Dublin. Royal Irish Academy.

SEBER, G.A.F. 1982 *The estimation of animal abundance and other parameters.* London. Charles Griffin.

SKINNER, C., SKINNER, P. and HARRIS, S. 1991 An analysis of some factors affecting the current distribution of badger (*Meles meles*) setts in Essex. *Mammal Review* **21**, 51-65.

THORNTON, P.S. 1988 Density and distribution of badgers in south-west England - a predictive model. *Mammal Review* **18**, 11-23.

WIJNGAARDEN, A. VAN and PEPPEL, J. VAN DE 1964 The badger (*Meles meles*) in the Netherlands. *Lutra* **6**, 1-60.

In: Hayden, T.J. (ed.) 1993 *The badger*, 57. Dublin. Royal Irish Academy.

BADGER BACULA FROM SOUTH-WEST IRELAND

H. Whelton* and S.B. Power†**

* **Zoology Department, University College, Cork.**
** **Dept. of Agriculture and Food, Regional Veterinary Laboratory, Bishopstown, Cork.**

†To whom correspondence should be addressed.

POSTER ABSTRACT

Valid aging techniques are fundamental to population studies, as for example understanding the epidemiology of disease such as bovine tuberculosis in badgers. Increasing size corresponds to increasing age and is reflected in the length of the animal or some part thereof. The differences are less marked in older animals. A limited sample (89) of male badger (*Meles meles*) bacula were available for study. Badger bacula show marked changes in both size and shape correlated with the attainment of sexual maturity; therefore the bacula from these males were examined with respect to bacular morphology, weight and length. Based on these results, it was possible to divide male badgers into juveniles (animals in their first year of life) and into adults. This is in agreement with results found from other studies. The conformation of each baculum was studied and some irregularities were observed. As most badgers with irregular bacula were positive for tuberculosis, there may be some connection between the occurrence of tuberculosis and the unusual conformation of the baculum.

In: Hayden, T.J. (ed.) 1993 *The badger*, 58-63. Dublin. Royal Irish Academy.

CRANIOMETRICS AND AGE DETERMINATION OF THE EURASIAN BADGER (*MELES MELES* L.)

A. I. Il-Fituri and T. J. Hayden†

Mammal Research Group, Department of Zoology, University College, Dublin

†To whom correspondence should be addressed.

INTRODUCTION

Although over 3000 badgers were culled in Ireland in an attempt to control TB in cattle between 1980 and 1989, few biological data were collected from these specimens (Dolan 1993). This situation was remedied somewhat in 1989 (Whelan and Hayden 1993). There are a number of reasons why the ages of badgers removed during control operations should be estimated. Examination of TB status and age may provide limited but valuable insights into the dynamics of the disease in the population. Comparison of the age structures of heavily infected with less infected populations may reveal the impact of TB on badgers. Examination of the age structure of the culled sample may reveal whether particular age classes are being missed, particularly if comparisons can be made with unculled populations (Cheeseman *et al.* 1987; Harris and Cresswell 1987). Finally, the effect of a removal operation on mobility and recolonisation may be examined with particular reference to the ages of badgers which recolonise an area (Cheeseman *et al.* 1993)

There are two broad approaches to the estimation of age in mammals. The first is to examine structures which may show periodic changes in growth rates, usually attributed to seasonal effects, and which exhibit discrete increments for each year of life. The second is to examine attributes which continuously increase or decrease during the life of the animal. Any method of age estimation should be calibrated with specimens of known age (Dapson 1980).

Ten age classes may be identified by a non-destructive examination of cranial and dental features, but most of these classes have been traversed by an animal within the first two years of life (Hancox 1988).

Sectioned teeth of badgers have been used in several studies in which it is assumed that annular structures in the cementum of canine or incisor teeth represent annual changes in the rate of tissue growth (see Ahnlund 1976; Hancox 1988 and references therein). However, in all these studies a proportion of the specimens (16-20%) had not formed legible annuli. Furthermore, in two separate but geographically close populations (Gloucester and Bristol) it has been concluded that there is a poor relationship between development of annuli and age and that annuli are rather poorly defined (Harris, unpublished; Cheeseman, unpublished). Similar findings were reported for Cornwall but not for Scotland (Fargher and Morris 1975). Although no Irish badgers of known age were available to us, it was decided (1) to survey a sample of badgers from Ireland for annuli in the cementum pad of the canine and to examine the agreement between annular counts and other general correlates of age such as tooth wear or bacular size; and (2) to analyse, using powerful multivariate techniques, the growth of the skull of known-age badgers from two British populations to determine whether there are subtle age-related changes in skull proportions not otherwise detectable and which might be suitable to discriminate between age classes.

METHODS

Irish badgers

Skulls and lower jaws of 123 Irish badgers were examined. These were drawn at random from a larger collection being assembled in the Department of Zoology, UCD. There were 54 males, 52 females and 17 of unknown sex. They were collected in Cavan (50) and Offaly (73).

British badgers

Access was obtained to 54 skulls of known-age badgers from Britain. Thirty-one (18 males and 13 females) were from Gloucestershire while 22 (13 males and 9 females) were from Bristol (Cheeseman *et al.* 1987; Harris and Cresswell 1987).There were 2 cubs and 2 yearlings; the remainder ranged from 2 to 10 years of age.

Craniometric measurements

A series of 11 standard measurements of the skull and 6 of the upper incisors and canines were taken from each intact specimen from Britain and Ireland (Wiig 1986). Data were collected from other specimens as damage allowed.

Annuli in cementum

Annuli in the cementum of the lower canines were determined for 106 Irish specimens only. The fragments of mandibles with the canine *in situ* were decalcified in 5% nitric acid, and sections (20-25μm) were cut with a clinical CO_2 microtome and stained with haematoxylin.

Analysis

The data were entered into a database on a Macintosh computer and analysed using the Statview 512™ and Data Desk software packages. The relationship between each cranial measurement and age or annular count in cementum were examined. Furthermore, a number of ratios of pairs of cranial measurements were examined. A number of curve-fitting routines were used to fit Von Bertalannfy, polynomial or logarithmic functions. In addition, a multivariate analysis (Principal Component Analysis) of the craniometric data from the British badgers was carried out to search for more subtle changes in skull growth. For principal components which were significantly affected by age, the relationship between age and component was examined. In addition, a canonical variate and a cluster analysis were used to examine if there was any clustering of specimens in relation to age.

RESULTS

Annuli in cementum of Irish badgers

Twelve badgers (11%) failed to develop annuli. Of the remainder, 83 (78% of the total) exhibited annuli which were considered relatively easy to interpret. The remaining 11 (11%) required more prolonged consideration and more repeated counting before arriving at the final figure. If animals adjudged by other criteria to be aged two years or less were excluded, then the proportion of adults for which no estimate could be made declined to 5%. In general, as mammals age, dentine is progressively deposited at the perimeter of the pulp cavity of teeth, leading to a progressive obliteration of the cavity. In Irish badgers there is an inverse relationship between the dimensions of the pulp cavity of canines and the number of cementum annuli in the same tooth (Fig. 1). This is the type of relationship that would be expected if annuli are deposited annually.

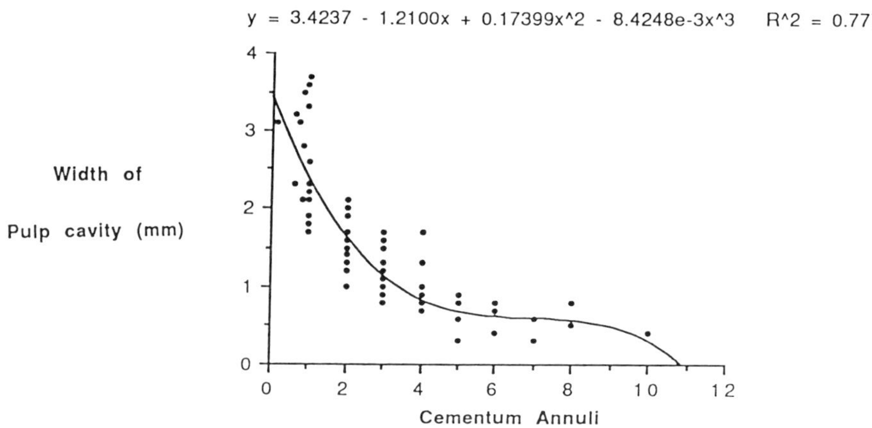

Fig. 1. Relationship between the width of the pulp cavity and the number of annuli in the cementum of lower canine teeth of badgers of unknown age from Ireland.

Morphometrics of British badgers

Bivariate plots of cranial dimensions against age indicated that the growth of the skull was essentially complete by the third year of age and that the variation was such that little additional information could be obtained beyond that which might be inferred from an inspection of the state of closure of cranial sutures (Fig. 2). The dimensions of the skulls were quite variable and the proportion of the variation in each variable which could be ascribed to age ranged from 10% to 56%. Thus age classes could not be identified. A number of ratios were then examined for an age effect (Fig. 3). Use of ratios did not improve identification of age classes, therefore a multivariate principal component analysis of 10 cranial measurements from the 34 badgers of known age was carried out. The first principal component, (PC_1 : overall size measure and increases with all variables except post-orbital constriction) accounted for almost 70% of the variation and the first five for over 95%. Only one principal component (PC_3 : related to skull breadth and increases with post-orbital constriction and decreases with post-orbital breadth) was

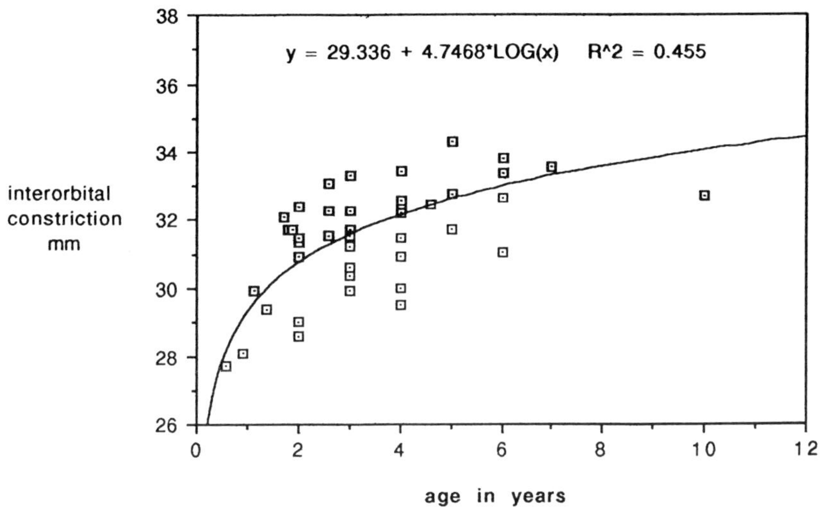

Fig. 2. Relationship between the distance across interorbital constriction and age of badgers from Gloucestershire and Bristol, UK.

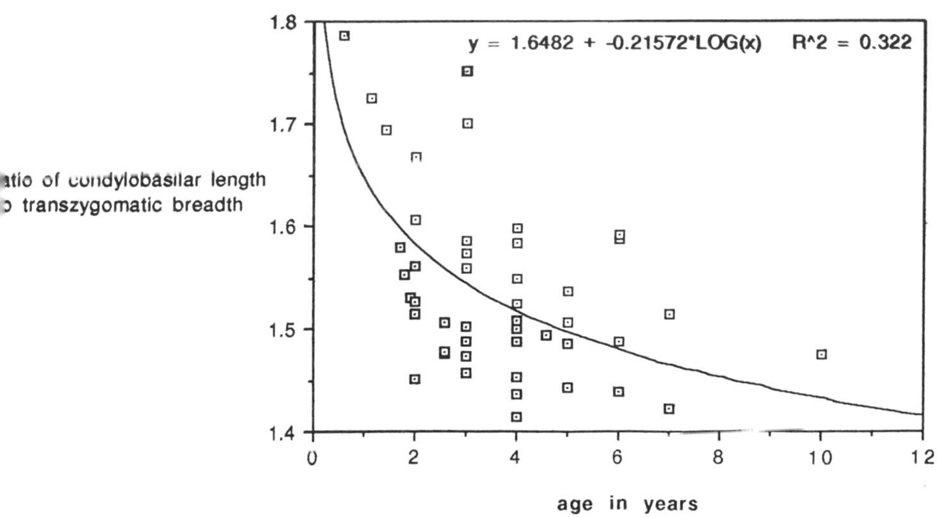

Fig. 3. Relationship between the ratio of condylobasilar length to transzygomatic breadth and age of badgers from Gloucestershire and Bristol, UK.

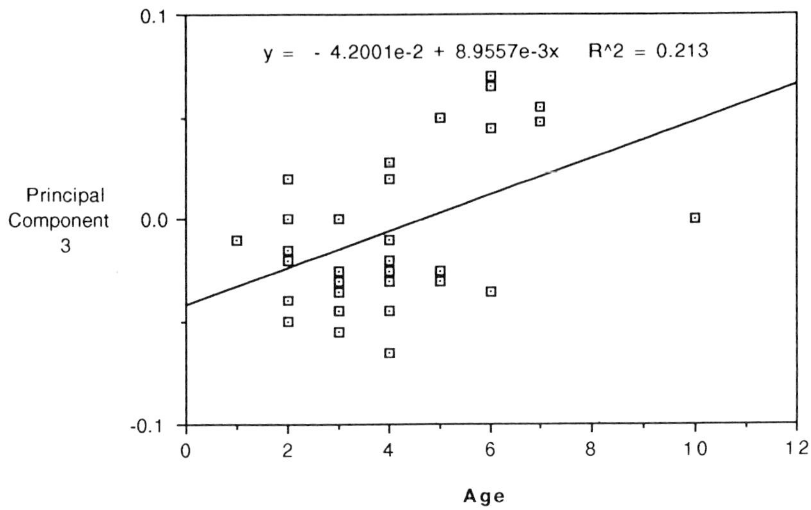

Fig. 4. Relationship between PC_3 and age of badgers from Gloucestershire and Bristol, UK.

significantly related to age ($P < 0.001$) But although PC_3 increased with age, the overlap between age classes is too great to distinguish between adjacent classes (Fig. 4). Neither could a canonical variate analysis or a cluster analysis accomplish adequate discrimination between age classes.

DISCUSSION

It is concluded that the variation in standard craniometric measurements of adult badgers contains an age-related component insufficient for reliably estimating the age of individuals older than three years of age. Therefore if estimates of the ages of badgers being culled in Ireland are to be obtained, the use of annuli in cementum must be considered. Almost 90% of the badgers examined had developed annuli. In male badgers the dimensions of the baculum in those with a single annulus were smaller than those of animals with two or more and were consistent with the dimensions given for yearlings in other studies (Fargher and Morris 1975). Furthermore, all animals with unfused cranial sutures had two annuli or less, while the most heavily worn teeth tended to contain more annuli, but this was somewhat variable. Thus while there was a tendency for the most worn teeth and those with the most occluded pulp cavities to occur in animals with greater numbers of annuli, it must nevertheless be stressed that this is a correlation between two variables both of which are assumed but not proven to be correlated with age. However, overall it may be concluded that cementum annuli could be used to age badgers in Ireland when the method is validated with a sample of badgers whose ages are known. It is therefore recommended that an attempt be made to accumulate a sample of jaws from individuals of known age to validate this method for this country, and to collect as a matter of routine lower jaws from culled badgers in the future. This will allow a more detailed exploration of the epidemiology of TB in badgers.

ACKNOWLEDGEMENTS

We are indebted to Dr S. Harris and Dr C. Cheeseman for providing access to known-age specimens. We thank the staff of the Department of Agriculture and Food, who provided the Irish specimens, and J.M. Lynch for some statistical advice. A. Il-F. thanks the Government of the Jamahirya of Libya for its support.

REFERENCES

AHNLUND, H. 1976 Age determination in the European badger (*Meles meles*). *Zeitschrift für Säugetierkunde* **41**, 119-25.

CHEESEMAN, C.L., MALLINSON, P.J., RYAN, J. and WILESMITH, J.W. 1993 Recolonisation by badgers in Gloucestershire. In T.J. Hayden (ed.), *The badger,* 78-93. Dublin. Royal Irish Academy.

CHEESEMAN, C.L., WILESMITH, J.W., RYAN, J. and MALLINSON, P.J. 1987 Badger population dynamics in a high density area. *Symposia of the Zoological Society of London* **58**, 279-94.

DAPSON, R.W. 1980 Guidelines for statistical usage in age estimation techniques. *Journal of Wildlife Management* **44**, 541-8.

DOLAN L. 1993 Badgers and bovine tuberculosis in Ireland: a review. In T.J. Hayden (ed.), *The badger,* 108-16. Dublin. Royal Irish Academy.

FARGHER, S. and MORRIS, P. 1975 An investigation into methods of age determination in the badger (*Meles meles*). London University unpublished report.

HANCOX, M. 1988 A review of age determination criteria in the Eurasian badger. *Lynx (Praha)* **24**, 77-86.

HARRIS, S. and CRESSWELL, W. J. 1987 Dynamics of a suburban badger (*Meles meles*) population. *Symposium of the Zoological Society of London* **58**, 295-311.

WHELAN, R. and HAYDEN, T. J. 1993 The reproductive cycle of the female badger (*Meles meles* L.) in east Offaly. In T.J. Hayden (ed.), *The badger,* 64-77. Dublin. Royal Irish Academy.

WIIG, Ø. 1986 Sexual dimorphism in the skull of minks (*Mustela vison*), badgers (*Meles meles*) and otters (*Lutra lutra*). *Zoological Journal of the Linnean Society* **87**, 163-79.

In: Hayden, T.J. (ed.) 1993 *The badger*, 64-77. Dublin. Royal Irish Academy.

THE REPRODUCTIVE CYCLE OF THE FEMALE BADGER (*MELES MELES* L.) IN EAST OFFALY

R.Whelan and T.J. Hayden[†]

Mammal Research Group, Department of Zoology, University College, Dublin

[†]To whom correspondence should be addressed.

ABSTRACT

Little is known about the reproductive cycle of the female badger in Ireland. Reproductive tracts and mammary glands of 548 adult female badgers, taken from Offaly between January 1989 and December 1990, were examined. Most sows in Offaly usually mate in late March/early April. Between 80% and 90% of the females mate, between 65% and 70% achieve successful implantation and between 35% and 40% exhibit copious lactation. Reproductive performance of yearlings is inferior at all stages to that of older females. Litter size *in utero* ranged from 1 to 4. Most of the cubs are born between the last week of January and the last week of February.

INTRODUCTION

While there have been several studies on the reproductive biology of the badger in other parts of Europe (Neal and Harrison 1958; Harrison 1963; Canivec 1966; Ahnlund 1980; Wandeler and Graf 1982a; Cresswell and Harris 1992), very little research has been carried out on Irish badgers.

Over recent years interest in the role of density-dependent processes in controlling the reproductive rate of badgers has increased. In particular it has been suggested that most (though not all) of the surviving cubs in a social group are the offspring of the more dominant sow in the clan (Kruuk 1978). In certain populations, however, some clans may produce two or three litters per year (Cheeseman *et al.* 1987; Harris and Cresswell 1987). Furthermore, there is not a clear relationship between the number of cubs produced and the number of adult females in a clan. Several studies have indicated that in some populations as few as 40% of cubs survive to their second year of life (Cheeseman *et al.* 1987). In areas of low density, however, over 80% of adult females have been identified as lactating (Neal 1986; Anderson and Trewhella 1985). While the benefits which accrue to a fertile female badger which does not breed remain to be satisfactorily demonstrated, the mechanism by which reproduction is suppressed is more amenable to investigation.

Reduction of female reproductive success might be accomplished at various stages of the reproductive process, by suppression of ovulation, by accelerated loss of blastocysts during the pre-implantation period, by lower rates of implantation, by increased resorption of fetuses, by inhibition of lactation post-partum, abandonment of neonates, or by infanticide/cannibalism of neonates by other members of the clan. This phenomenon, by whatever mechanism it is achieved, implies a considerable reserve reproductive capacity in badger populations.

This has important implications for the evaluation of control programmes to reduce badger numbers on the assumption that TB incidence in bovines will subsequently decline. It has been demonstrated that where badger populations are eradicated it may take 10 years for the original densities to be restored by immigration of founder animals from the periphery (Cheeseman *et al.* 1993). The duration of the recovery period can be expected to depend on the extent of the area to be recolonised, the population densities at the periphery and the original density to be restored. Nevertheless it has been argued/implied that the local depression in a population after a removal operation will persist sufficiently to establish a correlation between lower badger densities and lower incidence of TB in bovines (O'Connor and O'Malley 1989). What is less clear, however, is the efficacy of a control programme which, although it may remove the majority of a badger population, leaves substantial numbers of social groups intact (that is, containing at least one breeding pair). In such situations, if reproductive success is density-related, restoration of the original badger density may occur considerably sooner than expected, the ecological impact of the trapping programme may be minimal in the medium term and the likelihood of establishing an involvement of badgers in the aetiology of bovine TB, if such exists, might be rendered more difficult, if not impossible.

In 1987, the Department of Agriculture and Food decided to initiate a badger control operation in east Offaly on a scale never before attempted in this country (Eves 1993). Given that at the time there was no information available on badger densities, territoriality or reproduction for this country, it was proposed to collect information on these subjects, so that the impact of the impending programme might be better understood (O'Corry-Crowe *et al.* 1993). Furthermore, since removal of the whole population over a 600km^2 area was to be attempted, the usual caveats attached to the analysis, post mortem, of unrepresentative samples were unlikely to be a major problem (Harris and Cresswell 1987). We here provide a preliminary report on the reproductive status of female badgers taken in east Offaly from 1989 to 1990.

METHODS

Source of animals

For a number of years, the Department of Agriculture and Food has cleared areas of badgers where they were considered to be the main cause of an outbreak of TB. For an account of the criteria used to indicate a likely involvement of badgers see Flanagan (1993). Until recently no biological studies were carried out on these animals. In 1987 it was decided to subject the animals from one of these schemes to further examination. When the clearance operation was extended to other parts of Offaly in 1989, it was decided to continue the research.

Badgers were taken under licence by the Department of Agriculture and Food during a badger control operation as part of an investigation of bovine TB in an area of east

Offaly during 1989 and 1990. For each animal, capture location and body weight were available and samples, e.g. heads, reproductive tracts and mammary glands, were collected at regional veterinary laboratories where examination of the animals was carried out. The trapping programme was progressive across the study area and most of the areas were trapped on two occasions; thus animals can be subdivided into those taken from a first or second trapping session in both 1989 and 1990 (Table 1). The total number of adult females taken during the period 1989/90 is shown in Table 2.

Table 1. Animals caught in 1989 and 1990 in east Offaly.

		YEARLINGS	2+	TOTAL
1989	Trapping 1	88	239	327
	Trapping 2	8	33	41
1990	Trapping 1	9	20	29
	Trapping 2	47	104	151
TOTAL		152	396	548

Table 2. The number of animals per month caught in east Offaly during 1989 and 1990.

	JAN	FEB	MAR	APR	MAY	JUN	JUL	AUG	SEP	OCT	NOV	DEC
89	56	86	50	42	40	5	-	11	25	20	25	8
90	-	39	41	23	15	3	-	7	9	19	17	7

Age determination

Although annular structures can be readily identified in sections of decalcified canine teeth in over 70% of these specimens, there is no general agreement as to whether these can be used reliably to determine the age of the animals. This is compounded by the lack of a series of canine teeth for badgers of known age for comparison (Il-Fituri and Hayden 1993). Therefore, for the present purposes, badgers were conservatively separated into three age classes: cubs (animals in their first year of life), yearlings (animals in their second year of life) and those aged two years or older (animals in their third year of life or older, 2+).

Reproductive data

Reproductive tracts were examined and the presence and number of blastocysts, implantation sites (placental scars) and fetuses were noted. The weight and sex of each fetus was recorded. The zonal diameters of blastocysts flattened under a glass slide were measured using a microscope fitted with a micrometer eyepiece (Ahnlund 1980). Mammary glands were examined for evidence of lactation.

Badgers examined

The animals examined during 1989 and 1990 are shown in Table 1. A total of 368 sexually mature females were examined in 1989, of which 96 were yearlings, and a total of 180 were examined in 1990, of which 56 were yearlings. Trapping was carried out in all months except July, but the trapping effort was concentrated in autumn and spring. The numbers of animals examined in each month are shown in Table 2.

RESULTS

Births

The proportions of females which were pregnant (containing blastocysts or fetuses) throughout 1989 and 1990 are shown in Fig. 1. In the first two weeks of January in 1989, 66% of the adult females were pregnant. None of the remainder showed any macroscopic evidence of implantation. The first parous animal (exhibiting placental scars and lactation) was taken on 20 January (week 3). By 28 January (week 5), 31% of these pregnant females had given birth. The last pregnant animal was taken on 14 February. Most of the births occurred between 22 January and 4 February (i.e. between weeks 3 and 7). In 1990 about the same proportion of animals (63%) were pregnant overall as in 1989. No pregnant animals were taken after 28 February. Most of the births took place later than in 1989, between 18 and 28 February (weeks 4 and 10).

Overall, 43 females bearing fetuses were examined, 25 in 1989 and 18 in 1990. The mean litter size, assessed as the number of fetuses per female, was 2.9 $\pm$ 0.27 (range 1-4). There was no significant difference between years (2.8 $\pm$ 0.23 in 1989 and 3.3 $\pm$ 0.42 in 1990). In 1989, the first blastocyst was found on 2 February in a non-parous sow, but no others were found until 22 March. Given that the presence of blastocysts is indicative of mating, then most of the matings in 1989 occurred between 17 and 31 March (weeks 10 - 12). In 1990, the first blastocyst was found on 7 February but the second was not found until 5 April. Most of the matings occurred between 8 and 21 April (weeks 13 - 15). In both years between 60% and 65% of the females had mated by the end of April (week 16) (Fig. 1). Although fewer animals were taken in the summer months there was little further change until August, when the proportion of animals with blastocysts began to increase and reached 80-90% by the end of September (week 40, Fig. 1).

Age-related fertility

The fertility of yearlings is lower than older age classes. The proportion of yearlings with blastocysts rises to about 30-35% by early May (week 17), with little change thereafter until October when it reaches 60% (Fig. 2). The peak of matings for yearlings, which takes place in the last 2 weeks of April and the first week of May (weeks 15-17), is later than that of the 2+ animals (Fig. 2). It can be seen also that the increase in mated females which occurs in summer and autumn is mainly due to yearlings mating for the first time. There was no difference between the seasonal patterns shown by yearlings in 1989 and 1990. Furthermore, no yearlings taken between February and May showed evidence of either implantation or lactation. In this study 15 cubs were examined. None of these contained blastocysts.

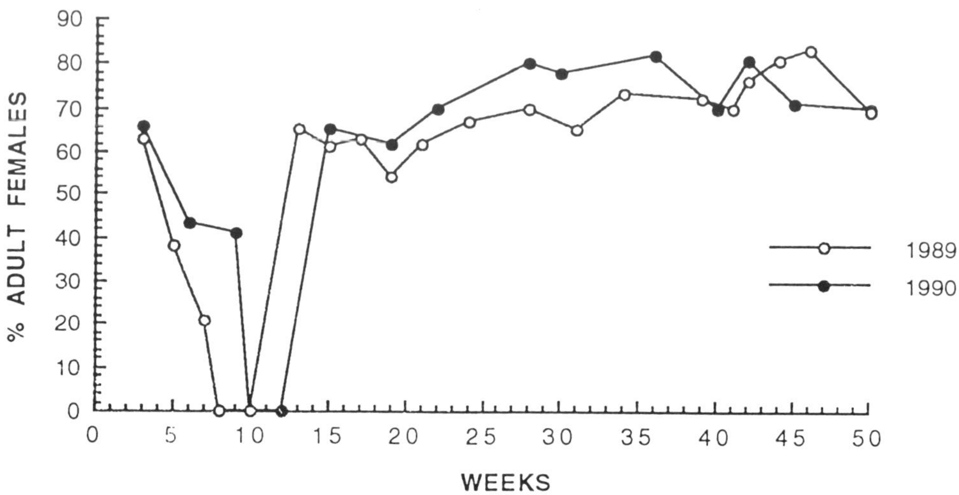

Fig. 1. The peak periods of births and matings for badgers in County Offaly during 1989 and 1990. This was found from the proportion of adult females in the population with fetuses or blastocysts.

Number of blastocysts per female

The number of blastocysts per mated female ranged from 1 to 5 (mean 2.66 $\pm$ 0.07). In yearlings between 1 and 3 were found (mean 2.41 $\pm$ 0.04); in older females the values ranged from 1 to 5 (mean 2.7 $\pm$ 0.2).

Blastocyst size

The zonal diameter for individual blastocysts taken throughout the year is shown in Fig. 3. Blastocysts increased in size during the period of delayed implantation. At the end of April, when most of the matings had occurred, the mean zonal diameter of the blastocyst was 0.6 $\pm$ 0.22mm (c.v. = 15.55%). By implantation this had increased to 3.57 $\pm$ 0.16mm (c.v. = 23.64%). The variation in size of blastocysts increased during the second half of the year. This was most pronounced from September onwards and was due to an effect of females with unusually small blastocysts (Fig. 3). This coincided with the autumnal increase in the proportion of animals with blastocysts (Fig. 1). The degree of variation within individual animals was similar throughout the year. When females taken between weeks 35 and 51 (September to December) are considered, 14% of these had blastocysts of two size classes, one typical of the whole sample for that time of year, the other substantially smaller.

Implantation

There was a loss of some blastocysts at implantation. The proportion of animals with blastocysts decreased from between 80% and 90% just prior to implantation to around

65-70% afterwards. This loss was particularly apparent in the yearlings, dropping from around 65% pre-implantation to about 35% thereafter. Once implantation has occurred (between the second week in December and the first week in January), the gestation period lasts about 50 days. The mean litter size for both years was 2.9. Lactation was first detected in early to mid-February and continued until at least the second week in June (the last lactating animal was taken on 12 June). The cubs reach maturity after their first year, at approximately 14 months.

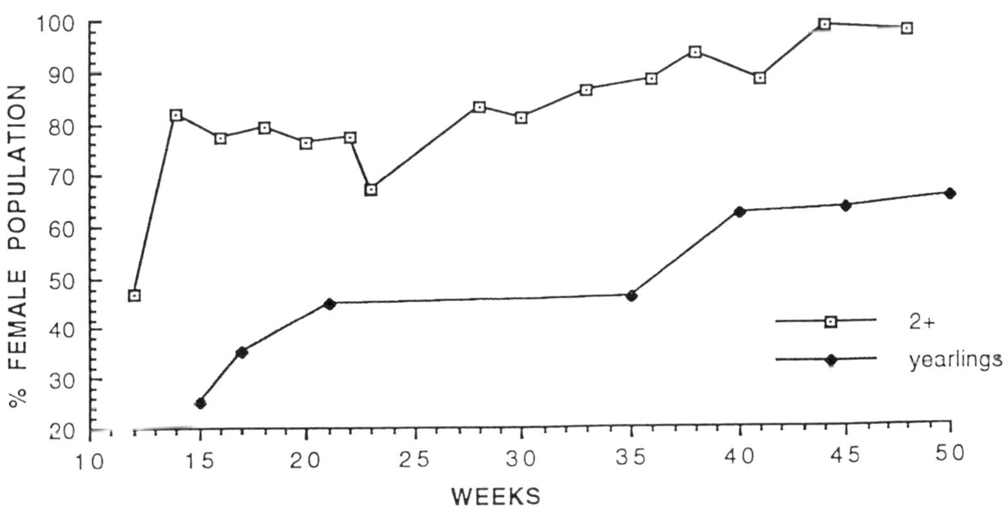

Fig. 2. Comparison of the mating success of yearling females and older animals. Mating success was judged on the proportion of each age class with blastocysts.

Previous reproductive history

An attempt was made to assess the effect of previous reproductive status on current reproduction. The presence of placental scars or evidence of lactation were taken as indicators of a previous reproductive attempt. Figure 4 shows the persistence of placental scars post-partum. Between 65% and 70% of females had implanted fetuses in late January. The incidence of placental scars was identical through May until August. Thereafter it became progressively more difficult to determine whether a female had borne fetuses the previous spring. It was decided, therefore, to assess the previous reproductive history of females taken earlier than 31 May. The probability of fertile matings occurring in consecutive years was over 0.8. Placental scars could be detected in 72% of females (2+) with blastocysts taken between 20 April and 31 May. This indicates that most of the adult females mate in consecutive years.

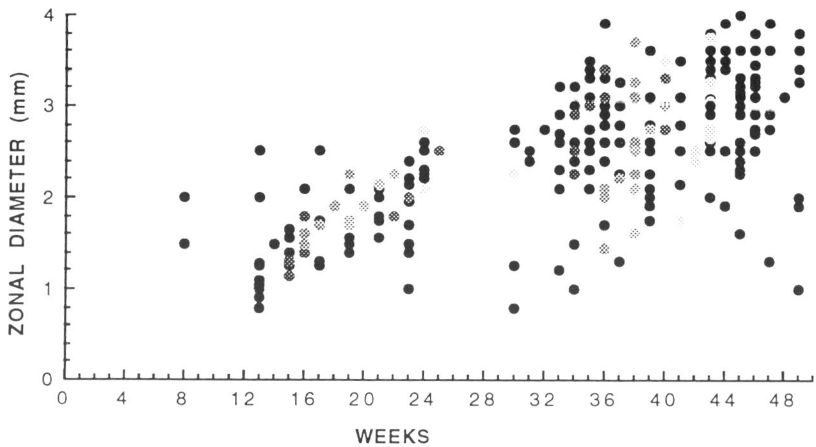

Fig. 3. Blastocyst growth throughout the year (1989/1990 combined). Each point may represent more than one blastocyst . N=418.

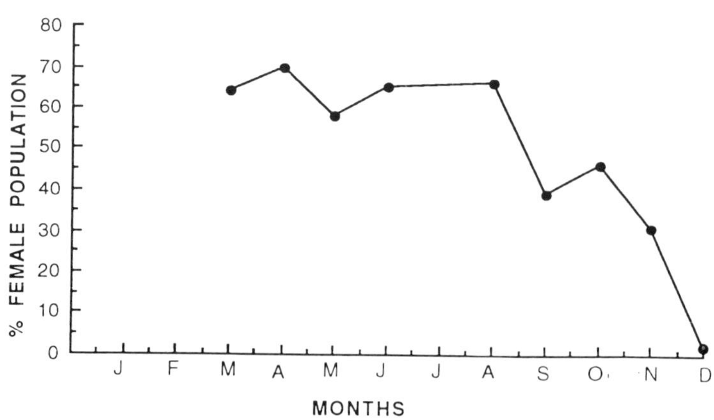

Fig. 4. Proportion of the adult females in the population with placental scars throughout the year (1989/1990 combined).

The changing proportions of adult females in various reproductive states over the first five months of the year are shown in Fig. 5. Parous females (those with placental scars) increased between weeks 3 and 9; most of the births occurred between weeks 3 and 7. From week 5 onward, lactating females were detected. This increased up to week 11. Between weeks 11 and 15 from 30-36% of adult females were lactating; thereafter the proportion of females whose mammary glands contained milk steadily declined. Thus of those females which contained fetuses, between 50% and 60% later showed evidence of copious lactation. The major spring increase in the proportion of females with blastocysts occurred between weeks 11 and 19 (Fig. 5).

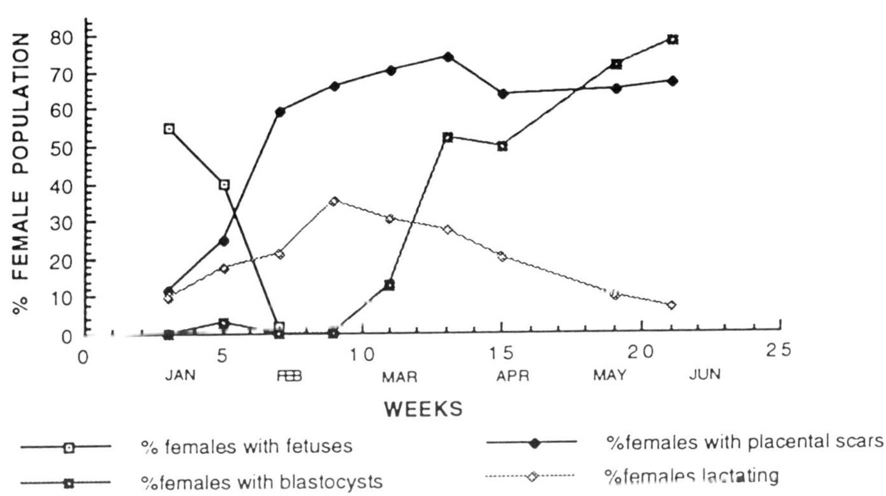

Fig. 5. The reproductive status of adult female badgers in County Offaly throughout January to June (1989/1990 combined).

DISCUSSION

Until recently there were rather few published data on the reproductive capacity of the European badger. An understanding of the allocation of reproductive effort in relation to lifetime reproductive success will only emerge from capture-mark-recapture studies (Harris and Cresswell 1987; Cheeseman *et al.* 1987). Such studies are of necessity long-term and require study populations within the spectrum of stable densities exhibited by badgers across their range. Useful information can, however, be derived from analysis of badgers post-mortem (Ahnlund 1980; Wandeler and Graf 1982a; Cresswell and Harris 1992; Cresswell *et al.*, in press), particularly if the sample is representative of the population in question. This may be achieved by collecting specimens from a variety of sources or by examining a single population. The present study is unusual in that it falls into the latter category. All the specimens came from within an area of 600km^2 and were taken over a 2-year period. The finding that most adult female badgers in County Offaly conceive each year is in general agreement with other studies which have analysed badgers post-partum (Ahnlund 1980; Wandeler and Graf 1982a; Cheeseman *et al.* 1987; Cresswell and Harris 1992).

Most adult females (90%) in the Offaly population contained blastocysts in late autumn. About 65-70% of these had acquired them by the end of May, the remainder by August. Both these periods coincide with seasons of enhanced territorial activity by males (Roper *et al.* 1986; Pigozzi 1990; O'Corry-Crowe *et al.* 1993). The variation in the size of the blastocysts increases from September onwards. This is consistent with an autumnal period of reproductive activity given that blastocysts increase in size during the pre-implantation period (Canivec 1966; Ahnlund 1980) and thus smaller blastocysts may be those most recently produced. Sows with small blastocysts were less frequently detected during the summer and those cases that were identified followed a pattern in agreement with what is known about mating dates for badgers (Ferris 1986; Neal 1986; Cresswell and Harris 1992).

By early January implantation had occurred, but only 65% of adult females examined in late January had either carried or currently bore fetuses. Thus only about 70-75% of the females which produce blastocysts progress to gestation proper. No cases of mortality *in utero* were detected over this period. It therefore seems likely that those females which did not breed failed to implant their blastocysts. In Cape porcupines, for example, females which have mated do not conceive when kept within their natal groups, and since this reproductive inhibition could not be explained by a failure to mate, it may have been caused by insufficient secretion of progesterone, necessary for the preparation of the uterus for implantation (Van Aarde and Van Wyk 1991).

During March, in Offaly, the proportion of adult female badgers with placental scars was equivalent to the proportion of adult females with fetuses detected in January. Placental scars are clearly visible until July-August but thereafter they become less obvious, and the proportion of animals with evidence of their previous gestation declines to zero by December. Thus it is extremely unlikely that scars from two successive pregnancies will be evident in the same animal. As the proportion of pregnant animals declines during the season of births, lactating females, as expected, begin to appear. The greatest proportion of lactating females (35%) was found in March and April. This represents slightly over half the proportion of females which established gestation. It is not possible to partition this perinatal loss between that represented by litters not brought to term and post-natal mortality. Other studies have indicated that between 35% and 40% of cubs born die before weaning (Cheeseman *et al.* 1987; Harris and Cresswell 1987), suggesting that this latter factor is more important. These studies also indicate that about 30-35% of adult females lactate each year. The proportion of lactating females in Offaly (35%) is in agreement with data obtained from field studies (Cheeseman *et al.* 1987; Harris and Cresswell 1987) and by the analysis of specimens post-mortem (Wandeler and Graf 1982a; Cresswell and Harris 1992).

The tentative conclusion that about 60% of females which properly established gestation then progressed to a full lactation feeding their own cubs cannot of course exclude the possibility of allomothering. We have not yet examined the reproductive status of individual females in relation to others from the same social group or sett but have noted, in passing, one instance of two lactating females from the same sett. It is not known whether they both had surviving cubs at the time.

Lactation in this study was assessed macroscopically by examining the mammary glands for milk. Some of the mammary glands were examined post-mortem in the veterinary laboratory itself. When samples of the glands were fixed for histological examination, lactating specimens could be readily identified by the cloudy appearance of the fixative caused by leakage of milk from the tissue. It is not as yet possible to identify the mechanism responsible for the early cub loss. Some females may fail to establish lactation. There are no data at present, however, on the histological changes which occur in the badger during mammogenesis, the involutionary changes which occur after the premature cessation of lactation, or on the normal process which occurs at weaning.

The litter sizes in east Offaly as assessed by the number of fetuses per female in late January ranged from 1 to 4 (mean 2.9 ± 0.27). This is within the range found in a number of studies reviewed by Anderson and Trewhella (1985). There was no evidence of a decrease from the mean number of blastocysts per sow detected the previous autumn. Thus loss of blastocysts at implantation appears to be largely an all or none process, in which females with smaller numbers of blastocysts may be more likely to lose them all. Furthermore, since the mean litter size as indicated by the number of fetuses in January is greater than the mean number of blastocysts (2.66) the previous autumn, it may be that the mated females least likely to achieve gestation are those with lesser numbers of blastocysts. The same cannot be said for survival of cubs since there is no information on the number of cubs being raised by individual sows.

A number of factors may affect the likelihood that a female will breed. These include age and body condition and size of social group. Breeding by cubs is rare (Ferris 1986; Cheeseman *et al.* 1987). Females are generally considered to become sexually mature in their second year (Neal and Harrison 1958; Ahnlund 1980; Wandeler and Graf 1982a) and are presumably capable of a fertile mating, implying that they would become primiparous on or about their second birthday. In two long-term studies on badgers, breeding, as assessed by the presence of palpable fetuses or lactation, first occurred in the third year (Cheeseman *et al.* 1987; Cresswell *et al.*, in press) or in the fourth year (Harris and Cresswell 1987). However, in these studies, it is, of course, not known what proportion of the various classes mated.

In this study the fertility of yearlings (year class 2) was lower than that of adults; 35% of yearling sows mated in the spring as opposed to about 80% of the older adults. By November 60% of yearlings had blastocysts, 90% of adults. Although 65-70% of mated adults progress to gestation, for those females mated as yearlings the proportion is 38%. This is slightly higher than the 20% found by Cresswell *et al.* (in press) but is still considerably lower than older adults. These are animals who would give birth during their third year of life. For this analysis we have assigned animals to three age classes, cubs, yearlings, and older animals (adults); nevertheless it seems that the lactation rate of sows which mated as yearlings is much lower than that of adults. None were found out of a sample of 14 examined between March and April. Thus over all stages of the reproductive process the performance of yearlings is inferior to that of older females. Little can be concluded regarding further effects of age on reproduction since we have not further subdivided the adults into component age classes. It may be that the proportion (10%) of adult badgers which do not mate each year is composed largely of the younger adults, i.e. those in the third year of life. Cresswell *et al.* (in press) noted a progressive increase in reproductive success with age in a sample of 352 female badgers collected from south-west England.

Assessing the age of a badger is not easy and a number of methods have been used with varying success (Ahnlund 1976; Wandeler and Graf 1982b; Hancox 1988). Although annuli which are laid down annually can be identified in sectioned teeth of badgers, this phenomenon is not universal in all populations. For example, badgers from Sweden and Switzerland may be readily aged by this technique (Ahnlund 1976; Wandeler and Graf 1982b) but it is not reliable for badgers from south-west England (Fargher and Morris, unpublished; Harris and Cresswell 1987). The situation in Ireland is still unresolved. About 70% of canine teeth from two populations of Irish badgers (Offaly: R. Whelan, unpublished; Cavan: Il-Fituri 1992) exhibit annuli and in general greater tooth wear is associated with higher numbers of annuli, but since we have as yet no access to known-age specimens from Ireland calibration of the method is not possible. Changes in skull shape with age have been used to estimate the age of some species of mustelids (Van Bree *et al.* 1966), but a multivariate analysis of skull shape of known-age badgers from Bristol and Gloucester failed to allow separation of age classes (Il-Fituri 1992). We have not yet assigned adult females to an age class and therefore cannot yet establish with certainty at what ages female badgers achieve the major component of their reproductive success.

In mammals nutrition and consequent body condition as indicated by body fat reserves are generally considered to regulate reproductive events, particularly puberty and ovulation, by mechanisms which are poorly understood (Bronson and Manning 1991; Suttie *et al.* 1991). Since most badgers in the Offaly population mate, then an effect of body condition, if it plays an important role in regulation of reproduction, might more likely act at implantation, as has been suggested from a consideration of body weights of badgers from Gloucestershire (Cheeseman *et al.* 1987), but Cresswell and Harris (1992) showed that females without blastocysts were characterised by smaller fat deposits and lower body weights. The relationship between body condition and the probability of implantation is less clear. We have not yet examined this factor but propose to do so in relation to social group size and habitat parameters which are available for the east Offaly area.

The temporal separation of ovulation/mating from implantation is of interest to an understanding of the evolution of mate choice (Sandell 1990) even if the neuroendocrine control of delayed implantation is still poorly understood (Mead 1989). It also permits separate analysis of the factors controlling respectively ovulation/oestrous and implantation. Photoperiod is the most important factor regulating and synchronising the termination of embryonic diapause and onset of implantation (Canivec and Bonnin 1979). In over half the badgers in Offaly mating occurs during an oestrous taking place within a month of parturition. What stimulates non-breeding females to mate at this time is not clear. It may be that in the badger there are a number of reproductive events controlled by photoperiod, some by decreasing and others by increasing day length.

Assuming that winter lethargy is widespread in the European badger (Fowler and Racey 1988) it might indicate that mating in December is unlikely, but this is not the case

(Neal 1986; Ferris 1986). Nevertheless the spring and autumn mating seasons may represent bet-hedging in relation to male choice by females. In the spring males might be selected on their ability to survive the worst period of the year; in autumn they might be selected on the basis of their ability to exploit the resources available during the season of abundance (Sandell 1990). It is not clear whether those adult females, in Offaly, which conceived in the autumn had previously mated unsuccessfully in the spring, and if so whether they use different criteria for mate choice in each season.

Individual females have been seen mating more than once in the same oestrous period or at intervals of up to 3 months and not always with the same male (Paget and Middleton 1974; Ferris 1986). For those females which subsequently give birth to cubs, it is not clear when their litters were conceived or if one or both males contributed to the paternity of the litter. We have observed unusual variance in blastocyst size in a small proportion (14%) of females examined in the autumn. This trend was also found in other studies (Cresswell *et al.,* in press, where 21.1% of the animals examined exhibited this type of super-foetation). It seems likely that this phenomenon represents the presence of two crops of blastocysts of different ages (Cresswell and Harris 1992). It remains to be established if both types of blastocysts are equally likely to implant successfully. The application of DNA fingerprinting techniques to determine the kinship of badger clans can be expected to reveal new insights into the consequences of variation in social organisation and reproductive patterns for badgers.

This study has established the principal features of the reproductive cycle of the female badger in Offaly: mainly vernal breeding, high fertility, loss of blastocysts by failure of implantation, and high perinatal mortality. It is thus similar to that established by field studies of other populations. Whether effects of factors such as age, body condition, social group size or habitat characteristics can be identified remains to be seen.

ACKNOWLEDGEMENTS

We gratefully acknowledge the assistance, both financial and technical, of the following in enabling us to carry out this project: the staff of the Department of Agriculture and Food, the District Veterinary Office (Tullamore), ERAD, Teagasc, the Wildlife Service, and the regional veterinary laboratories at Abbotstown, Athlone and Limerick.

REFERENCES

AHNLUND, H. 1976 Age determination in the European badger (*Meles meles* L.). *Zeitschrift für Säugetierkunde* **40**, 119-25.

AHNLUND, H. 1980 Sexual maturity and breeding season of the badger (*Meles meles*) in Sweden. *Journal of Zoology* **190**, 77-95.

ANDERSON, R.M. and TREWHELLA, W. 1985 Population dynamics of the badger (*Meles meles*) and the epidemiology of bovine tuberculosis (*Mycobacterium bovis*). *Philosophical Transactions of the Royal Society of London* **310B**, 327-81.

BRONSON, F.H and MANNING, I.M. 1991 The energetic regulation of ovulation: a realistic role for body fat. *Biology of Reproduction* **44**, 945-50.

CANIVEC, R. 1966 A study of progestation in the European badger. *Symposium of the Zoological Society of London* **15**, 15-26.

CANIVEC, R. and BONNIN, M. 1979 Delayed implantation is under environmental control in the badger (*Meles meles* L.). *Nature* **278**, 849-50.

CHEESEMAN, C.L., MALLINSON, P.J., RYAN, J. and WILESMITH, J.W. 1993 Recolonisation by badgers in Gloucestershire. In T.J. Hayden (ed.), *The badger,* 78-93. Dublin. Royal Irish Academy.

CHEESEMAN, C.L., WILESMITH, J.W., RYAN, J. and MALLINSON, P.J. 1987 Badger population dynamics in a high density area. *Symposium of the Zoological Society of London* **58**, 279-94.

CRESSWELL, W.J. and HARRIS, S. 1992 To breed or not to breed: the control of reproduction in badger populations. *NERC News* (April 1992), 14-16.

CRESSWELL, W.J., HARRIS, S., CHEESEMAN, C.L. and MALLINSON, P.J. (in press) To breed or not to breed: an analysis of the social and density dependent constraints on the fecundity of female badgers (*Meles meles*). *Philosophical Transactions of the Royal Society of London.*

EVES, J. 1993 The East Offaly Badger Research Project : an interim report. In T.J. Hayden (ed.), *The badger,* 166-73. Dublin. Royal Irish Academy.

FERRIS, C. 1986 Mating season and early maturity of badgers in Kent. *Journal of Zoology, London* **209A,** 282.

FLANAGAN, P. 1993 Badgers and bovine tuberculosis: epidemiology and case studies. In T.J. Hayden (ed.), *The badger,* 131-8. Dublin. Royal Irish Academy.

FOWLER, P.A. and RACEY, P.A. 1988 Overwintering strategies of the badger (*Meles meles*) at 57°N. *Journal of Zoology, London* **214**, 635-51.

HANCOX, M. 1988 A review of age determination criteria in the Eurasian badger. *Lynx (Praha)* **24**, 77-86.

HARRIS, S. and CRESSWELL, W.J. 1987 Dynamics of a suburban badger (*Meles meles*) population. *Symposium of the Zoological Society of London* **58**, 295-311.

HARRISON, R. J. 1963 A comparison of factors involved in delayed implantation in badgers and seals in Great Britain. In A.C. Enders (ed.), *Delayed implantation,* 99-114. Chicago University Press.

IL-FITURI, A.I. 1992 An evaluation of indicators of age in Irish badgers. Unpublished M.Sc. thesis, National University of Ireland.

IL-FITURI, A.I. and HAYDEN T.J. 1993 Craniometrics and age determination in the Eurasian badger (*Meles meles* L.). In T.J. Hayden (ed.), *The badger,* 58-63. Dublin. Royal Irish Academy.

KRUUK, H. 1978 Spatial organisation and territorial behaviour of the European badger (*Meles meles*). *Journal of Zoology, London* **184**, 1-19.

MEAD, R.A. 1989 The physiology and evolution of delayed implantation in carnivores. In I.L. Gittleman (ed.), *Carnivore behaviour, ecology and evolution*, 437-64. Cornell University Press.

NEAL, E. 1986 *The natural history of badgers.* London. Croom Helm.

NEAL, E. and HARRISON, R.J. 1958 Reproduction in the European badger (*Meles meles*). *Transactions of the Zoological Society of London* **29**, 67-131.

O'CONNOR, R. and O'MALLEY, E. 1989 *Badgers and bovine tuberculosis in Ireland.* Dublin. Economic and Social Research Institute/ERAD.

O'CORRY-CROWE, G., EVES, J. and HAYDEN, T.J. 1993 Sett distribution, territory size and population density of badgers (*Meles meles* L.) in east Offaly. In T.J. Hayden (ed.), *The badger,* 35-56. Dublin. Royal Irish Academy.

PAGET, R.I. and MIDDLETON, A.L.V. 1974 Some observations on the sexual activities of badgers (*Meles meles*) in Yorkshire in the months December to April. *Journal of Zoology, London* **173**, 256-60.

PIGOZZ1, G. 1990 Latrine use and function of territoriality in the European badger (*Meles meles*) in a Mediterranean coastal habitat. *Animal Behaviour* **39**, 1001-2.

ROPER, T.J., SHEPHERDSON, D.J. and DAVIES, J.M. 1986 Scent-marking with faeces and anal secretion in the European badger (*Meles meles*): seasonal and spatial characteristics of latrine use in relation to territoriality. *Behaviour* **97**, 94-117.

SANDELL, M. 1990 The evolution of seasonal delayed implantation. *Quarterly Review of Biology* **65**, 23-42.

SUTTIE, J. M., FOSTER, D.L., VEANVLIEK, B.A., MORLEY, T.R. and CORSON, I.D. 1991 Influence of food intake but independence of body weight on puberty in female sheep. *Journal of Reproduction and Fertility* **92**, 33-9.

VAN AARDE, R.I. and VAN WYK, V. 1991 Reproductive inhibition in the Cape porcupine (*Hystrix africaeaustralis*). *Journal of Reproduction and Fertility* **92**, 13-19.

VAN BREE, P.J.H., JENSEN, B. and KLEIN, L.J.K. 1966 Skull dimensions and the length/weight relation of thc baculum as age indicators in the common otter. *Danish Review of Game Biology* **47**, 89-104.

WANDELER, R.I. and GRAF, M. 1982a The reproductive cycle of female badgers (*Meles meles* L.). *Revue Suisse de Biologie* **89**, 1009-16 (in German, English summary).

WANDELER, R.I. and GRAF, M. 1982b Age determination in the European badger (*Meles meles L.*). *Revue Suisse de Biologie* **89**, 1017-23.

In: Hayden, T.J. (ed.) 1993 *The badger*, 78-93. Dublin. Royal Irish Academy.

RECOLONISATION BY BADGERS IN GLOUCESTERSHIRE

C. L. Cheeseman,*† P. J. Mallinson,* J. Ryan and J.W. Wilesmith****

*** Central Science Laboratory, Ministry of Agriculture, Fisheries and Food, Tangley Place, Worplesdon, Guildford, Surrey GU3 31A.**
**** Epidemiology Unit, Central Veterinary Laboratory, New Haw, Weybridge, Surrey KT15 3NB.**

†To whom correspondence should be addressed.

ABSTRACT

The process by which badgers recolonise cleared areas is of central importance in the formulation of strategies designed to control bovine tuberculosis in badgers and cattle. Recolonisation was investigated as part of a long-term study of badgers and bovine tuberculosis in Gloucestershire. Following the complete clearance of defined areas of this infected badger population, it took nine to ten years for the population to recover to the pre-removal density. During the initial phase of recolonisation, social organisation was disrupted and badgers travelled over greater than usual distances. There was evidence of reduced mortality and higher than normal weight gain by badgers during the recovery phase in comparison to an undisturbed area. In the post-removal period tuberculosis infection was re-established in the badger population.

INTRODUCTION

In Britain and Ireland the badger has been identified as a potential reservoir of bovine tuberculosis (*Mycobacterium bovis*) and the species has been implicated in the transmission of the disease to cattle (Muirhead *et al.* 1974; Downey 1990). As a result, control measures have been undertaken in both countries in an attempt to limit the spread of *M. bovis* infection from badgers to cattle. The process by which badgers recolonise cleared areas after control measures have been implemented is of direct importance to the formulation of the control strategy itself. How soon do immigrant animals move in? Is there a compensatory increase in fertility? How long does it take the population to recover to the carrying capacity? Is there a risk that immigrant badgers will introduce infection? Answers to these questions are needed in order to determine the optimal frequency and intensity of control operations, and indeed to assess the usefulness of control measures in the long term.

There are no published long-term studies of badger recolonisation. Anderson and

Trewhella (1985), using simulation models of badger population dynamics, predicted that recovery time would be approximately five years. This estimate was based on recovery time being calculated as l/r, where r (the intrinsic population growth rate for badgers) was taken as 0.2. Obviously this estimate is very crude and takes no account of the various biological factors which will influence recovery time. These may include (1) the size of the cleared area, (2) the proportion of the population removed, (3) the carrying capacity of the habitat and its consequent influence on badger population dynamics, (4) badger population density around the cleared area, and (5) any population pressure in the control area post-removal. Given these additional constraints, it is likely that the estimate of five years will represent an optimal recovery time.

Of particular interest in the development of strategies to control *M. bovis* infection in badgers and cattle is the behaviour of badgers in the area around a control zone, and whether neighbouring social groups of badgers expand their territories to fill the 'void'. In a study of carnivore social behaviour, Kruuk and Macdonald (1985) described badgers as a 'contractor species', living in social groups which occupy the smallest economically defensible area, increasing their group size within this area if resources allow. If this is so, it might be predicted that the removal of a number of social groups in an area would not result in the expansion of neighbouring social group territories.

A long-term prospective study of badger population dynamics and the epidemiology of TB in wild badgers has taken place in south-west England (Cheeseman, Wilesmith *et al.* 1987; 1988). In order to investigate some of the questions posed above, the opportunity was taken to study badger recolonisation when the two outbreaks of *M. bovis* infection in cattle occurred in this area in 1978 and 1979. Trapping operations were undertaken to remove all badgers from the areas where the outbreaks occurred (Cheeseman, Jones *et al.* 1981). Recolonisation of the two cleared areas was monitored up to the end of 1990 and this paper reports the findings of this work.

METHODS

Study areas

The main study area lies in the Cotswold escarpment region of Gloucestershire in south-west England, where the incidence of *M. bovis* infection in cattle has been consistently much higher than the average for Britain (Wilesmith 1983). The area supports a high-density badger population with more than 20 adults/km^2 (Cheeseman, Jones *et al.* 1981; Cheeseman, Wilesmith *et al.* 1987). A description of the study area is given by Cheeseman, Jones *et al.* (1981). The two areas within the main study area where badger removal took place are known as Atcombe and North Woodchester, comprising five and six social groups respectively. Their badger populations were completely removed in 1978 and 1979 respectively. After the initial clearance, natural recolonisation was permitted and there was no further disturbance to badgers in either area, apart from routine trapping and other activities associated with the investigation.

Determination of social group territories

Bait-marking was used to delineate social group territories using the method described by Kruuk (1978) and Cheeseman, Jones *et al.* (1981). This was carried out annually, in the spring of each year, prior to removal in 1978 and 1979 and in each subsequent year up to and including 1990. The information from bait-marking was combined with field signs, such as obvious badger paths, to gain an impression of the spatial organisation of badger territories. The mean distance of active latrines from associated main setts, with 95 per cent confidence intervals, as assessed by bait-marking, provided a measure of the extent of movements of badgers within the removal areas.

Methods of capture and radio-tracking

An attempt was made to catch and radio-track badgers which moved into the cleared areas. Methods of catching and radio-tracking are described in Cheeseman and Mallinson (1979). For the purpose of illustrating the ranges covered by individual badgers, a convex polygon was drawn around the outermost points of location of the badger over the period of time it was observed.

A capture-mark-recapture programme provided data for the study of the dynamics and social structure of the population, and the movements of badgers within the study area. Initially the capture-mark-recapture regime involved a rotation of trapping effort throughout the study area over the year, so that each main sett was trapped two or three times in a year. In 1981 this continued, but greater effort was devoted to social groups known to contain individuals infected with *M. bovis*. Subsequently, in October 1987, the study area was subdivided into five trapping sectors and each was trapped in rotation once every three months, resulting in each main sett being trapped four times each year.

At first capture, badgers were tattooed on the abdomen and both ears were tagged (Cheeseman and Harris 1982), and their sex and year of birth, if a cub or yearling, were recorded. In cases where it was not obvious whether a badger was a yearling or an adult it was placed in the two- to three-year-old category for the purpose of the analyses. Each badger was weighed, its general condition, degree of tooth wear and reproduction state, if female, noted, and its site of capture recorded.

On each occasion that individual badgers were caught, clinical samples were taken and examined for the presence of *M. bovis*. Methods of sampling and laboratory screening are described in Cheeseman, Wilesmith *et al.* (1988).

Analyses

Badgers caught during the period 1979 to 1990 in the two removal areas were allocated to one of three categories in each year. These were:

(i) animals (cubs) born in the area in the year, i.e. first caught as a cub in the area with no evidence of immigration;

(ii) immigrant animals in the year, i.e. caught in the previous year(s) outside the area or in the case of adults and yearlings not caught previously (NB: some of these adults and yearlings may have been resident but evaded capture in previous years);

(iii) resident adult and yearling animals in the area, i.e. caught in the area in the previous year(s).

If an individual was not captured for one or more years between captures in the area, and if there was no evidence of any movement, it was assumed to have been present in the area in the intervening years.

The same analysis was conducted on the capture records for nine contiguous social groups in the south-western part of the main study area to provide a comparison with an undisturbed sector of the population.

Annual mortality rates from 1980 to 1990 were estimated as follows. An individual's lifespan was taken to be from the year of birth to the year of death. In the case of animals not found dead, the year of death was assumed to be the year in which it was last caught. These rates were determined for all age classes of animals by sex for the two removal areas combined and for the remainder of the study area which was undisturbed.

The mean weights, with 95 per cent confidence intervals, for cubs, yearlings and adults were calculated in three-monthly time periods for records accumulated during the period 1978/9 (depending on the removal area) to 1990 inclusive, for badgers in the two removal areas combined and for the remainder of the study area. Badgers caught at one sett in the North Woodchester removal area were excluded from this analysis. This was because these badgers were fed significant quantities of food on a daily basis by a member of the public resident in the study area.

RESULTS

Social group territories

The results of bait-marking in the Atcombe and North Woodchester areas for 1978 (prior to removal), 1981 (2-3 years post-removal) and 1988 (9-10 years post-removal) are shown in Figs 1, 2 and 3 respectively. Prior to removal (Fig. 1), territorial boundaries were clear-cut with a mean territory size of 23.3ha. Figure 2 illustrates that in the first few years post-removal, all main setts were reoccupied, badger movements overlapped considerably and there were no obvious territories. Over subsequent years territories gradually became apparent, and by 1988 (Fig. 3) the picture was stable once more with a pattern of territorial organisation markedly similar to that seen prior to removal. One difference was the emergence of one territory in North Woodchester in an area which supported two territories prior to removal. Mean territory size in 1988 increased slightly to 25.6ha.

This pattern of clear-cut territories pre-removal, disruption post-removal and a gradual return to clear-cut territories over a period of years was reflected in the mean distance of latrines with bait-markers from main setts over the years 1978-1990 (Fig. 4). This distance increased sharply during the first 1-2 years after removal had taken place, with a gradual return to the pre-removal level after approximately 5-6 years.

The ranges of two adult female badgers which moved into the removal areas during the initial phase of recolonisation are shown in Fig. 5. Their range sizes were 92.5ha (B37Q) and 108.8ha (W69Q) and covered areas which overlapped many of the pre-removal territories. Both these individuals utilised several main setts during the period of observation and frequently changed sett.

Numbers of badgers present

The number of adult badgers (resident and immigrant) and cubs present in the Atcombe and North Woodchester removal areas is shown in Figs 6a and b. These data may be compared with similar data for the nine contiguous undisturbed social groups of badgers (Fig. 6c).

Fig. 1. Bait returns showing clearly defined territories in the two removal areas prior to removal. (N) = North Woodchester sett; (A) = Atcombe sett; o = latrine; straight lines connect main setts to latrines where marked bait was recovered.

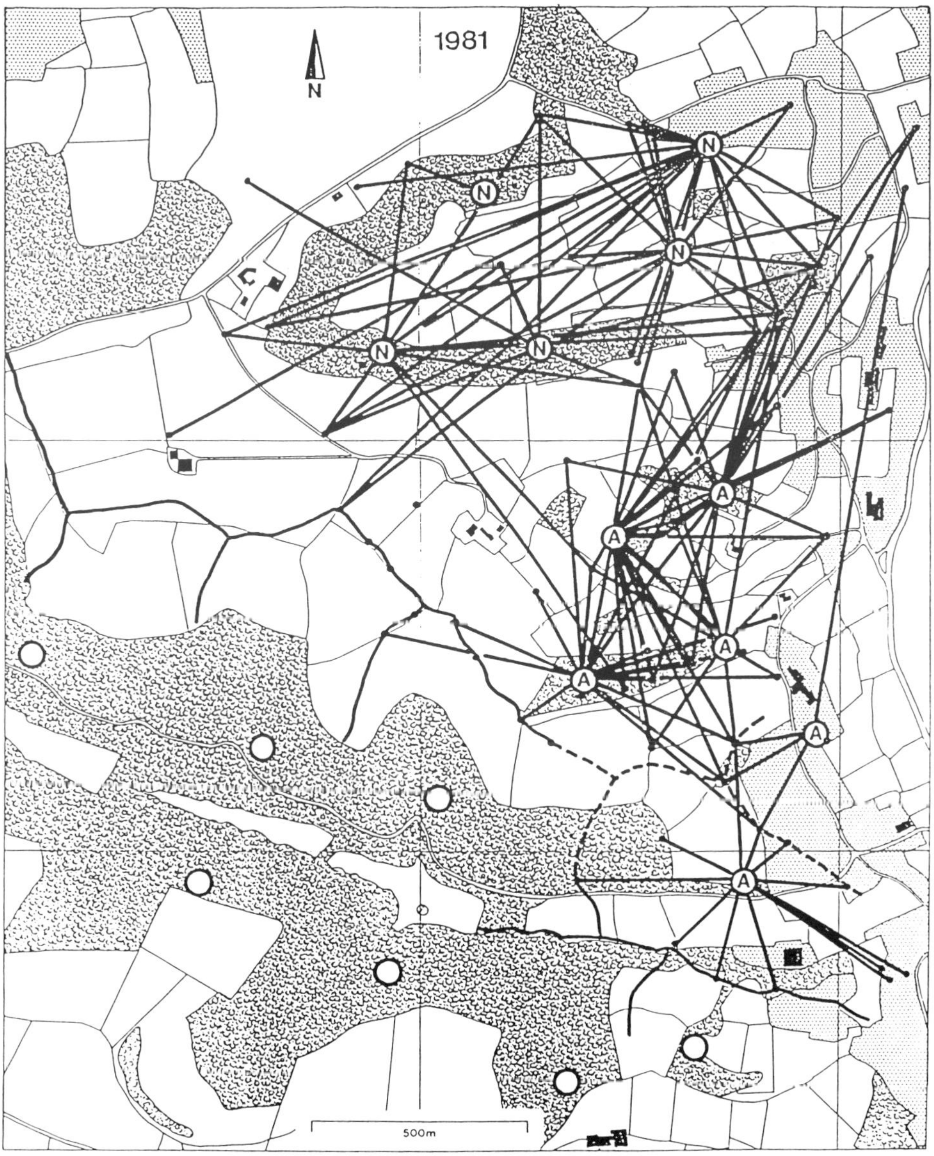

Fig. 2. Bait returns showing lack of territorial organisation in the two removal areas two and three years post-removal. (N) = North Woodchester sett; (A) = Atcombe sett; o = latrine; straight lines connect main setts to latrines where marked bait was recovered.

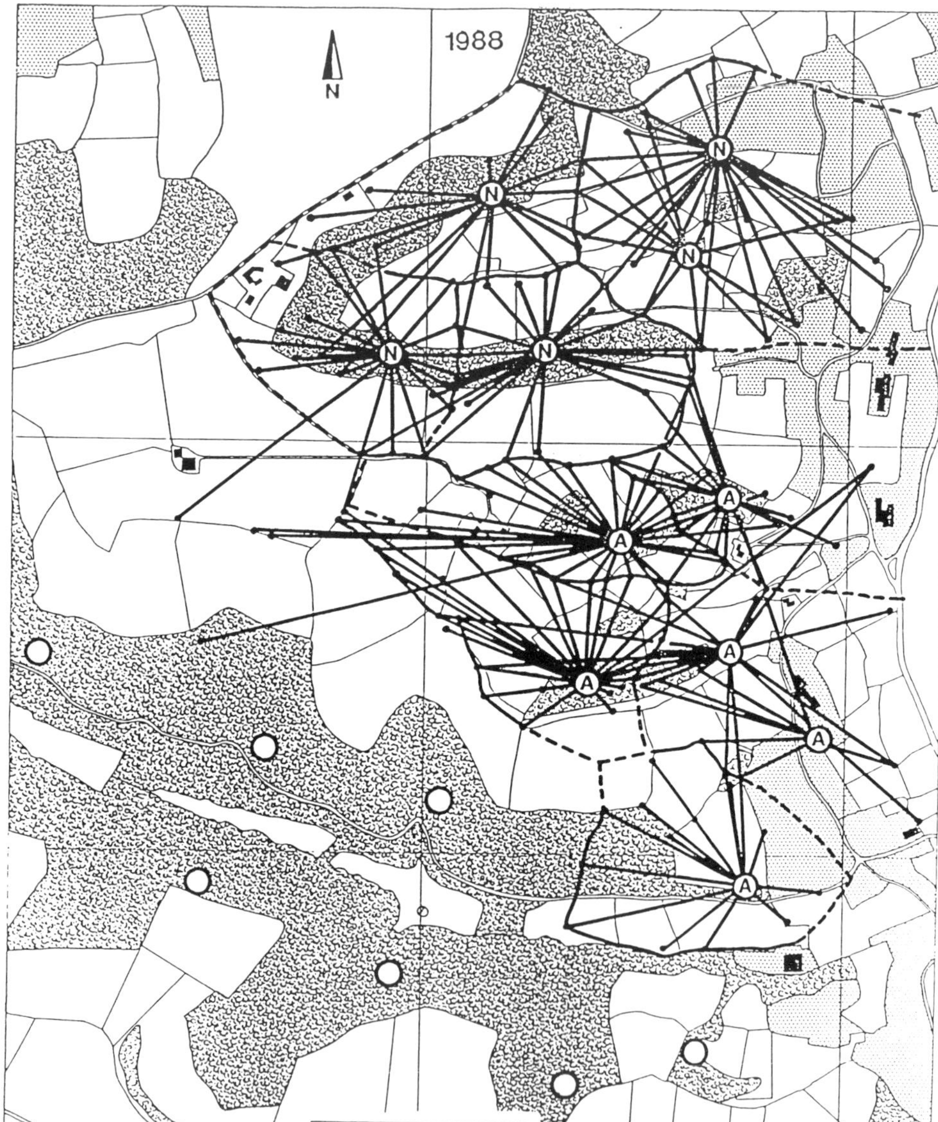

Fig. 3. Bait returns showing clearly defined re-established territories in the two removal areas nine and ten years post-removal. (N) = North Woodchester sett; (A) = Atcombe sett; o = latrine; straight lines connect main setts to latrines where marked bait was recovered.

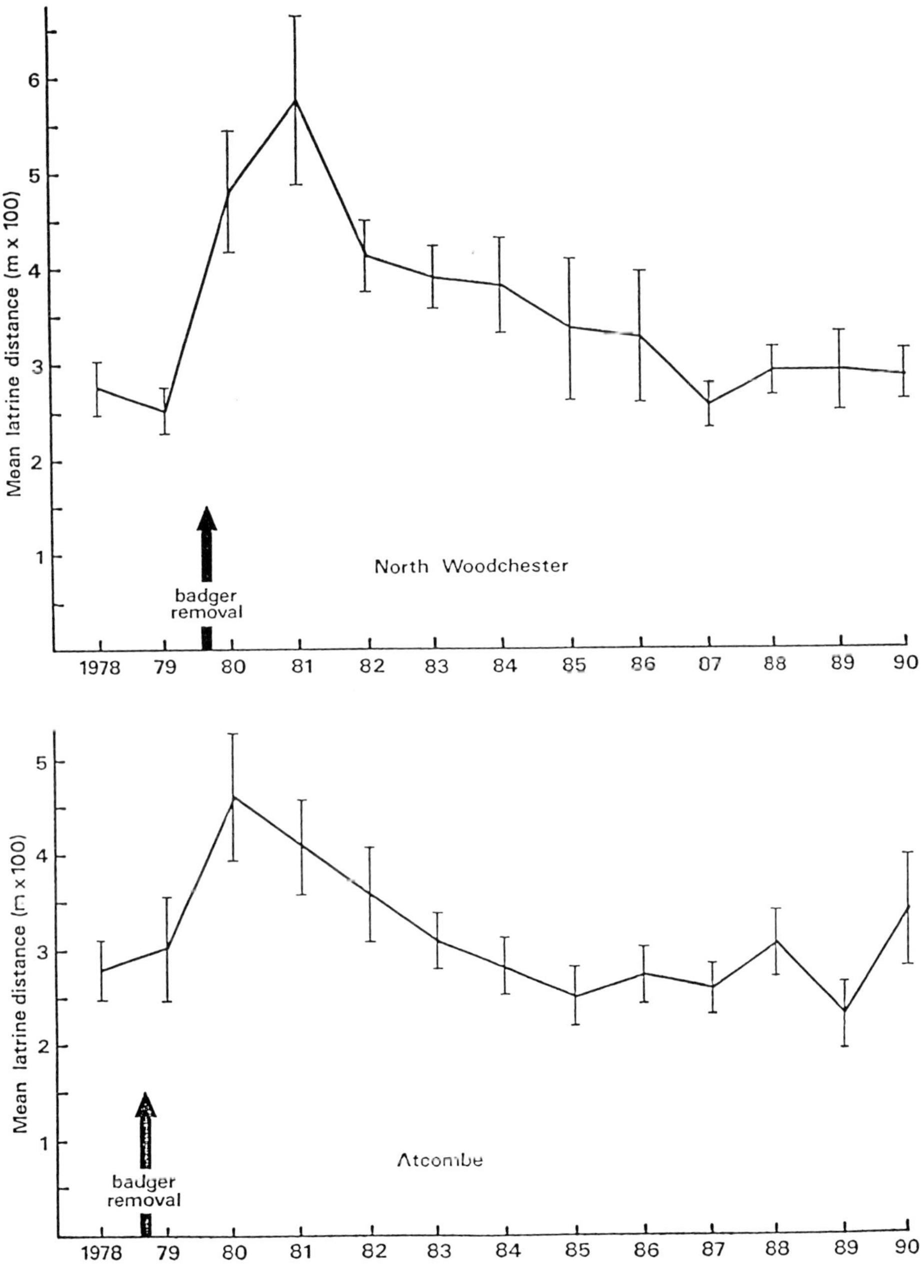

Fig. 4. Changes in the mean distance of latrines from main setts for the two removal areas 1978-1990.

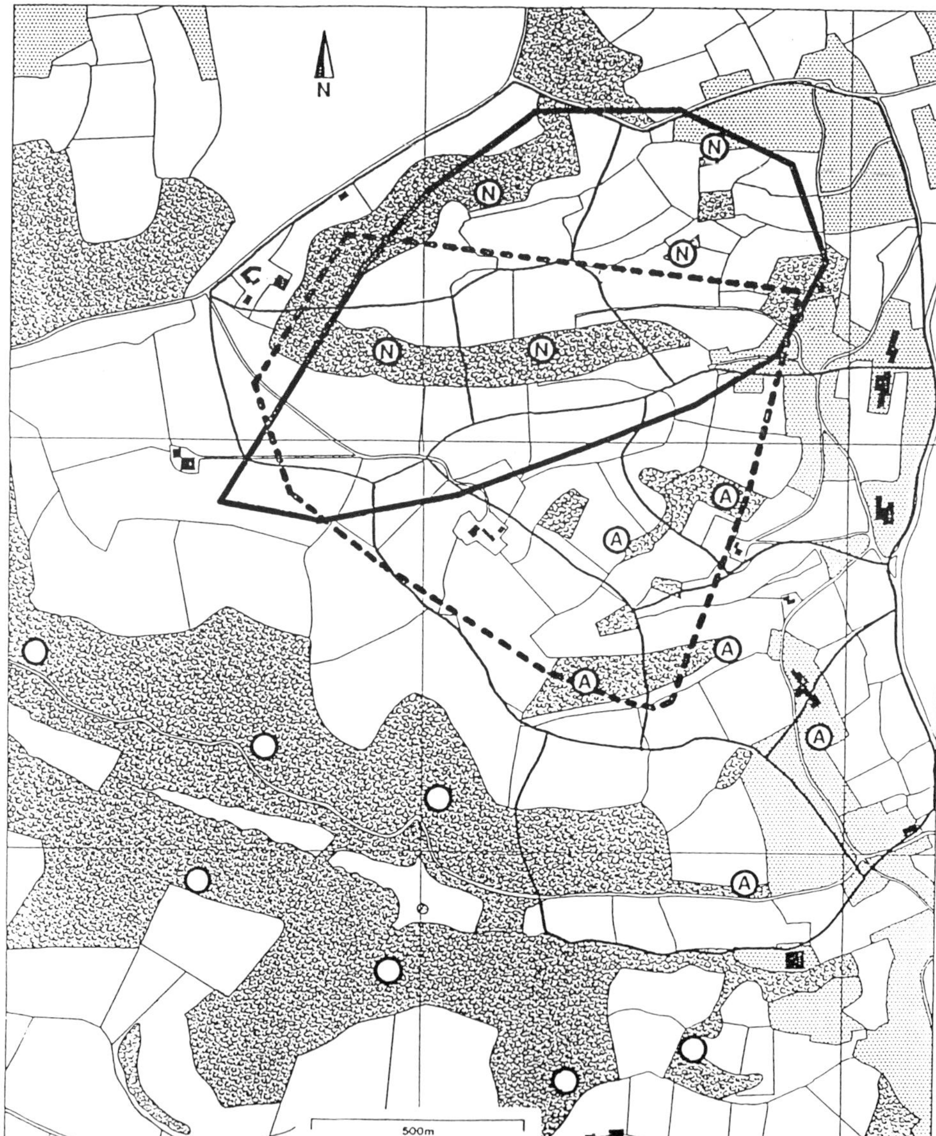

Fig. 5. Ranges of two badgers radio-tracked in the North Woodchester area post-removal. —— Badger B37 (female) tracked 30/11/79 - 22/1/81; ---- Badger W69 (female) tracked 5/12/79 - 26/2/81; —— pre-removal social group territory boundaries; (N) = North Woodchester sett; (A) Atcombe sett.

In the Atcombe removal area (Fig. 6a) it was not until 1988, ten years post-removal, that the number of adult badgers present reached a level similar to that seen prior to removal. The population stayed at a similar level in 1989 and 1990 (it should be noted that the 1990 estimates represent a minimum number of badgers present, and will be subject to revision when data for 1991 become available). The proportion of cubs to resident badgers present in the Atcombe removal area was highly variable over the first few years post-removal, and by 1988, 1989 and 1990 was similar to that seen in the nine undisturbed social groups. A different pattern of population recovery was seen in the North Woodchester removal area (Fig. 6b). Very few adult resident badgers were present from 1980 through to 1985. There was then a steady increase and by 1989, ten years post-removal, the adult population had reached the pre-removal level. No cubs were present in 1980 and 1981. By 1989 and 1990 cub numbers, as a proportion of adults present, were similar to the Atcombe area and the nine undisturbed social groups. In the nine undisturbed groups (Fig. 6c) adult badger numbers showed a steady increase from 1979 to 1989. The proportion of cubs was high in the early years, reached a low in 1982, and from 1983 to 1990 remained relatively stable at a mean adult : cub ratio of 2.33:1. The number of immigrant badgers exceeded the number of resident badgers during four of the first five years post-removal in the North Woodchester area, but in the Atcombe area immigrants only exceeded residents in the first year post-removal. By 1988 onwards, the proportion of immigrants was low in both areas. In the nine undisturbed groups the proportion of immigrants was low in all years.

Mortality

Apparent mortality rates for male and female badgers in the removal areas and the nine undisturbed social groups are shown in Table 1. In the undisturbed areas the mean male mortality for the years 1980-1989 (32.1% per annum) was higher than that of females (22.8%) ($X^2 = 4.227$, $P<0.05$). This difference was not seen in the removal areas where male and female mortalilty was 22.0% and 22.5% per annum respectively ($X^2 =$ 20.015, $P>0.05$). The greater mortality of males in the undisturbed groups compared to the removal areas was not significant ($X^2 = 1.893$, $F>0.05$).

Seasonal weights of badgers

Analysis of the mean weights of badgers in the removal areas and the remainder of the study area (Table 2) revealed a consistently greater weight in the badgers within the removal areas except for yearlings in the October to December quarter. The most marked difference was between cubs caught in the second quarter of the year (April to June), cubs in the removal areas having a greater mean weight than cubs in the undisturbed social groups.

Evidence of M. bovis *infection in badgers*

Five badgers infected with *M. bovis* were identified in the Atcombe and North Woodchester areas during the post-removal period (Table 3). The clinical samples taken from these individuals indicated a respiratory route of infection in all cases.

(A)

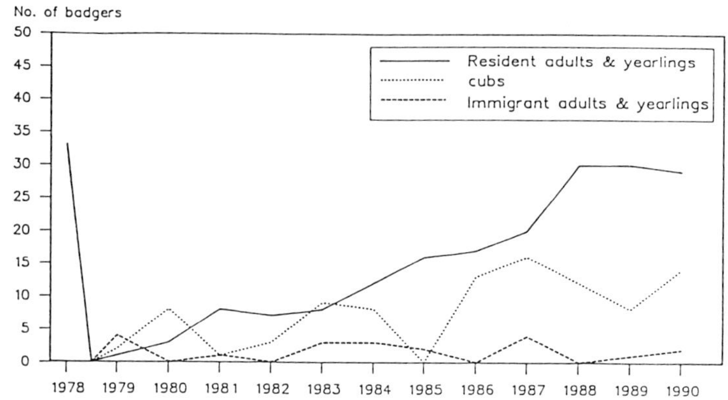

(B)

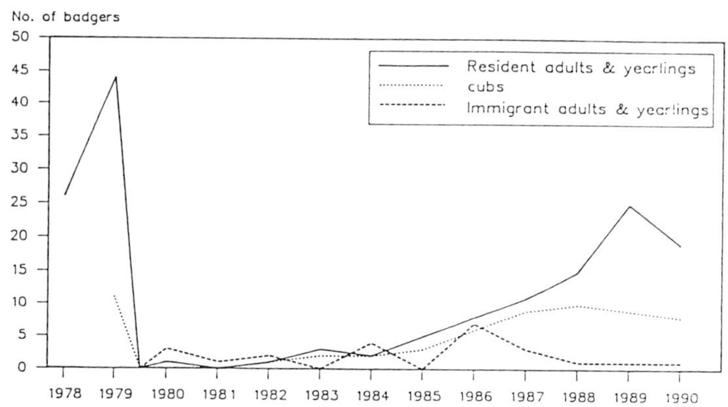

(C)

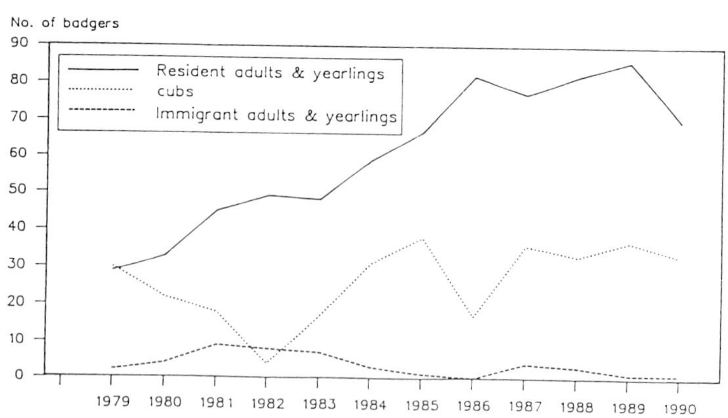

Fig. 6. Numbers of badgers present in: (A) the Atcombe removal area 1978-1990; (B) the North Woodchester removal area 1978-1990; (C) the nine contiguous undisturbed social groups.

Table 1. Apparent mortality rates for known-age badgers in the removal areas (A) and the undisturbed areas (B).

(A)

	TOTAL POPULATION			APPARENT DEATHS			APPARENT MORTALITY (%)		
	M	F	TOT	M	F	TOT	M	F	TOT
1977	0	0	0	0	0	0	0.0	0.0	0.0
1978	1	0	1	0	0	0	0.0	0.0	0.0
1979	1	3	4	0	0	0	0.0	0.0	0.0
1980	5	8	13	3	1	4	60.0	12.5	30.8
1981	3	8	11	0	1	1	0.0	12.5	9.1
1982	4	11	15	0	2	2	0.0	18.2	13.3
1983	8	16	24	1	6	7	12.5	37.5	29.2
1984	12	15	27	3	2	5	25.0	13.3	18.5
1985	10	15	25	0	1	1	0.0	6.7	4.0
1986	19	24	43	6	3	9	31.6	12.5	20.9
1987	25	34	59	6	10	16	24.0	29.4	27.1
1988	29	37	66	4	9	13	13.8	24.3	19.7
1989	35	36	71	10	11	21	28.6	30.6	29.6
1980-89	150	204	354	33	46	79	22.0	22.5	22.3

(B)

	TOTAL POPULATION			APPARENT DEATHS			APPARENT MORTALITY (%)		
	M	F	TOT	M	F	TOT	M	F	TOT
1977	12	10	22	7	6	13	58.3	60	59.1
1978	21	17	38	15	6	21	71.4	35.3	55.3
1979	31	37	68	15	15	30	48.4	40.5	44.1
1980	34	39	73	15	7	22	44.1	17.9	30.1
1981	33	50	83	9	10	19	27.3	20	22.9
1982	28	53	81	8	13	21	28.6	24.5	25.9
1983	35	56	91	8	11	19	22.9	19.6	20.9
1984	52	72	124	19	16	35	35.5	22.2	28.2
1985	77	87	164	16	16	32	20.8	18.4	19.5
1986	79	95	174	22	21	43	27.8	22.1	24.7
1987	78	109	187	22	25	47	28.2	22.9	25.1
1988	85	111	196	30	28	58	35.3	25.2	29.6
1989	87	104	191	40	30	70	46.0	28.8	36.6
1980-89	588	776	1364	189	177	366	32.1	22.8	26.8

Table 2. Comparison of cub, yearling and adult weight (kg) in the removal areas and the study area.

			Jan-Mar	Apr-Jun	Jul-Sep	Oct-Dec
CUBS	STUDY AREA	MEAN		2.8	4.4	7.4
		95% CI		2.8-2.9	4.3-4.5	7.3-7.6
		N		345	531	334
	REMOVAL AREAS	MEAN		3.2	4.5	9.1
		95% CI		2.9-3.4	4.2-4.8	7.5-8.7
		N		46	82	32
YEARLINGS	STUDY AREA	MEAN	7.5	7.7	8.0	10.1
		95% CI	7.4-7.7	7.5-7.8	7.8-8.2	9.9-10.4
		N	211	215	210	131
	REMOVAL AREAS	MEAN	7.9	7.9	8.4	9.9
		95% CI	7.5-8.3	7.3-8.6	7.8-8.9	8.9-10.9
		N	27	20	34	16
ADULTS	STUDY AREA	MEAN	8.9	8.0	8.1	10.3
		95.% CI	8.8-9.1	7.9-8.1	8.0-8.2	10.2-10.5
		N	301	522	835	460
	REMOVAL AREAS	MEAN	9.2	8.1	8.3	10.6
		95% CI	8.7-9.7	7.6-8.7	8.0-8.6	10.1-11.1
		N	38	41	122	52

Table 3. Details of individual badgers found to be infected with *M. bovis* in the Atcombe and North Woodchester removal areas.

Badger ID	Sex	Age at first capture	Date of first capture	Dates known to be present	Date first known to be infected	Area found	
						Atcombe	North Woodchester
E61	M	Cub	05/07/88	05/07/88 - 21/03/89	21/03/89		*
E93	M	Cub	20/09/88	20/09/88 - 20/03/90	20/03/89		*
N53	F	Cub	31/07/84	31/07/87 - 21/02/89	21/03/89		*
O11	M	Cub	27/05/82	27/05/82 - 20/03/89	20/03/89	*	*
R48	M	Cub	20/06/78	04/09/79 - 28/11/89	16/11/88	*	

DISCUSSION

It is apparent that badger social organisation was considerably disrupted in the removal areas for several years post-removal, and that a stable picture did not emerge until 9-10 years had elapsed. Social group territories in the undisturbed part of this population have been found to be highly stable from year to year (Cheeseman and Mallinson 1979; MAFF 1990). Observations on the two badgers which were radio-tracked show that the removal operations resulted in increased mobility by recolonising badgers. In the undisturbed part of the study area female badgers typically range over an area not larger than 15-20ha (Cheeseman and Mallinson 1979). This represents less than 25% of the areas covered by the two radio-tracked females in the removal areas. Compared with badgers in the undisturbed areas, these individuals ranged over relatively long distances and used more main setts than usual. In the context of badgers being described as 'contractor species' (Kruuk and Macdonald 1985) it is interesting to note that the removal operations did not result in the extension of neighbouring badger social group territories to occupy the vacant habitat. The movements of immigrant badgers were confined within the removal areas, and there was no evidence of any extensive badger movement between the removal area and the neighbouring undisturbed social groups (Cheeseman, Wilesmith *et al.* 1988).

It is difficult to know whether the number of badgers present before removal represented the true carrying capacity of the area. In the undisturbed groups the population slowly increased over the period of study. This is believed to be due to the absence of any persecution in this population since the study began. With factors such as slight changes in the trapping routine over the period of study, which may have influenced the data, and relatively small sample sizes to compare in the removal and undisturbed areas, it is difficult to draw any firm conclusions on the demographic consequences of removal. It was hoped, for example, to compare breeding rates of females in the removal and undisturbed areas, but very few breeding females were caught in the removal areas in the first few years post-removal. It has been shown (W. J. Cresswell, personal communication) that the full reproductive capacity is not realised by badgers in high-density populations. Thus it might be expected that a greater fertility would be exhibited by badgers in the removal areas in the period during which the population was below carrying capacity.

Density-dependent mortality has been demonstrated in this population and there is some evidence of density-dependent fertility (Cheeseman, Wilesmith *et al.* 1987; MAFF 1989; 1990). It is clear, however, that the badger populations in the removal areas took several years to recover. The slightly slower rate of recovery at North Woodchester compared to Atcombe is probably due to the fact that the bulk of the Atcombe area lies between North Woodchester and the main undisturbed part of the study area. The study area as a whole is bordered by areas where routine badger removal operations have taken place as a result of outbreaks of *M. bovis* infection in cattle. The undisturbed part of the study area therefore probably provided the main source of immigrant animals, and the rate of recolonisation at North Woodchester would be expected to be slower owing to its more isolated situation. The estimated recovery time of five years (Anderson and Trewhella 1985) thus represents a minimum which will be influenced by environmental variables such as recolonising pressure. Badgers were totally removed from the two removal areas described in this investigation. If only partial population reduction was achieved during a control operation, recovery time would be expected to be more rapid because there would be no initial recolonisation phase.

The greater mortality of male badgers in the undisturbed social groups compared to the removal areas may be associated with the lack of territoriality in the removal areas during the recolonisation period. It has been recorded that male badgers play a greater role in the maintenance of territory than females (Cheeseman, Cresswell *et al.* 1988). A lowered level of territoriality would presumably lead to less movement by males and a reduced frequency of aggressive interactions.

The generally greater mean weight of badgers in the removal areas may be due to reduced competition for food. The significantly greater mean weight of cubs in the April-June quarter in the removal areas may be a reflection of the better condition of their mothers during lactation compared with their counterparts in the undisturbed groups.

The prevalence of *M. bovis* infection in the Atcombe and North Woodchester areas at the time of removal was 6.9% (2/29) and 31.6% (12/38) respectively (Cheeseman, Jones *et al.* 1981). It was not possible to estimate the true prevalence during the post-removal period to compare with the post-mortem findings at the time of removal, but the presence of infection in the badger population post-removal was clearly demonstrated by the discovery of five infected individuals, the first of which was detected in November 1988. None of the five infected badgers exhibited infected bite wounds which would indicate possible transmission of *M. bovis* infection by territorial aggression with the neighbouring undisturbed population. Because of the long period between removal operations and the detection of the first infected badger, it is considered unlikely that any of these badgers became infected from residual infection in the setts occupied by the original badger population. It is more likely that some of the recolonising badgers brought disease with them from the surrounding area where *M. bovis* infection is known to be present (Cheeseman, Wilesmith *et al.* 1988).

The rate of recolonisation and subsequent population recovery, and the risks the immigrant badgers carry in the reintroduction of infection, are key points in the formulation of badger control strategies. Bovine tuberculosis has been shown to be endemic in the British badger population and there is increasing evidence that the disease is also endemic in the Irish badger population (Downey 1990). This study has shown that badgers will recolonise areas cleared by control operations, albeit at a slow rate, and that *M. bovis* infection can be reintroduced during the process of recolonisation.

ACKNOWLEDGEMENTS

We are grateful to the farmers and landowners whose cooperation enabled this study to take place. David Handoll, John Howell and Paul Spyvee provided first-class support in the field throughout the study.

REFERENCES

ANDERSON, R.M. and TREWHELLA, W. 1985 Population dynamics of the badger

(*Meles meles*) and the epidemiology of bovine tuberculosis (*Mycobacterium bovis*). *Philosophical Transactions of the Royal Society* **310B**, 327-81.

CHEESEMAN, C.L. and HARRIS, S. 1982 Methods of marking badgers (*Meles meles*). *Journal of Zoology, London* **197**, 289-92.

CHEESEMAN, C.L. and MALLINSON, P.J. 1979 Radio-tracking in the study of bovine tuberculosis in badgers. In C.J. Amlaner and D.W. Macdonald (eds), *A handbook on biotelemetry and radio tracking,* 649-56. Oxford. Pergamon Press.

CHEESEMAN, C.L., CRESSWELL, W.J., HARRIS, S. and MALLINSON, P.J. 1988 Comparison of dispersal and other movements in two badger (*Meles meles*) populations. *Mammal Review* **18**, 51-9.

CHEESEMAN, C.L., JONES, G.W., GALLAGHER, J. and MALLINSON, P.J. 1981 The population structure, density and prevalence of tuberculosis (*Mycobacterium bovis*) in badgers (*Meles meles*) from four areas in south-west England. *Journal of Applied Ecology* **18**, 175-804.

CHEESEMAN, C.L., WILESMITH, J.W., RYAN, J. and MALLINSON, P.J. 1987 Badger population dynamics in a high density area. *Symposia of the Zoological Society of London* **58**, 279-94.

CHEESEMAN, C.L., WILESMITH, J.W., STUART, F.A. and MALLINSON, P.J. 1988 Dynamics of tuberculosis in a naturally infected badger population. *Mammal Review* **18**, 61-72.

DOWNEY, L. 1990 *Ireland's TB problem - what can and must be achieved.* Dublin. ICOS.

KRUUK, H. 1978 Spatial organisation and territorial behaviour of the European badger (*Meles meles*). *Journal of Zoology* **184**, 1-19.

KRUUK, H. and MACDONALD, D.W. 1985 Group territories of carnivores: empires and enclaves. In R.M. Sibley and R.H. Smith (eds), *Behavioural ecology: ecological consequences of adaptive behaviour,* 521-36. Oxford. Blackwell Scientific Publications.

MAFF 1989 *Bovine tuberculosis in badgers.* Thirteenth Report. London. HMSO.

MAFF 1990 *Bovine tuberculosis in badgers.* Fourteenth Report. London. HMSO.

MUIRHEAD, R.H., GALLAGHER, J. and BURN, K.J. 1974 Tuberculosis in wild badgers in Gloucestershire: epidemiology. *Veterinary Record* **95**, 552-5.

WILESMITH, J.W. 1983 Epidemiological features of bovine tuberculosis in cattle herds in Great Britain. *Journal of Hygiene, Cambridge* **90**, 159-176.

In: Hayden, T.J. (ed.) 1993 *The badger,* 94-107. Dublin. Royal Irish Academy.

MORPHOMETRIC AND GENETIC VARIATION AMONG BADGER POPULATIONS

J.M. Lynch,* G. O'Corry-Crowe,* C.L. Cheeseman, S. Harris,*** E.J. Duke* and T.J. Hayden*†**

* **Mammal Research Group, Department of Zoology, University College, Dublin.**
** **Ministry of Agriculture, Fisheries and Food, Worplesdon Laboratory, Tangley Place, Worplesdon, Guildford, Surrey GU3 3LQ, UK.**
*****Department of Zoology, University of Bristol, Woodland Road, Bristol BS8 IUG, UK.**

† To whom correspondence should be addressed.

ABSTRACT

There exists some confusion as to the means by which mammals colonised Ireland after the last Ice Age. A number of putative land-bridges have been proposed by various authors, but as yet no one has utilised multivariate techniques to examine the differences in body form between British and Irish mammals, with a view to testing hypotheses about the origins of Irish mammals. We here report the variation between several European and Irish badger populations using statistical examination of cranial measurements and polymorphisms of selected enzyme systems as revealed by electrophoretic techniques. The skulls of Irish, Scottish and English badgers are morphologically distinct. Furthermore, populations of badgers in Bristol, Gloucestershire, Offaly and Down were separable. A review of published and new research on the genetic structure of several populations of badgers reveals that genetic differences can be demonstrated. Possible factors which may contribute to these morphologic and genetic differences are discussed.

INTRODUCTION

The European badger (*Meles meles*) has a Palaearctic distribution, ranging from Japan in the east to Ireland in the west (Neal and Cheeseman 1991). Ireland represents the western limit of its range, where it forms an important part of a mammalian fauna less diverse than that of Britain or mainland Europe. This fauna is thus of considerable biogeographical interest, as most elements (including the badger) must have colonised the island (either naturally or otherwise) during the last 10,000 years. To this end, we are currently investigating the relevance of an analysis of skull structure to determine the probable origins of the Irish fauna. It has been concluded after consideration of craniometry that the taxonomy of the badger within its European range has not been adequately studied (Hell and Paule 1989).

It must, of course, be emphasised that the proper study of genetic relationships between populations requires an analysis of the genotype or the genetic material itself. However, if it can be established that craniometric variation is not due solely to environmental effects, or that morphological differences between populations are, at a first approximation, proportional to genetic distance, then cranial anatomy provides a convenient means of comparing mammalian populations, especially when analysed by multivariate statistical techniques. While craniometric analyses have been used for some time to provide information on relationships between populations (e.g. Manning 1971; Huson and Page 1980), there is some debate as to the relative importance of genetic and environmental influences on such patterns of variation (Atchley *et al.* 1981).

In this study we report the extent of craniometric variation in several badger populations from Britain and Ireland, and quantify the morphological differences between them. We further review evidence for genetic differentiation between badger populations and assess the degree to which there is concordance between genetic and morphologic variation. To our knowledge, this is the first usage of multivariate morphometric techniques in the analysis of variation between badger populations in Europe. Similarly, we know of no studies which compare genetic variation across badger populations in a systematic manner.

MATERIAL AND METHODS

Material

Skulls of badgers (n = 122) were available from a number of sites in Ireland and the United Kingdom (Table 1). The Offaly badgers were all taken during control operations carried out by the Department of Agriculture and Food, and were from the Killeigh region (Eves 1993). The approximate area of this site is 5.0km^2, and the density of badgers ranges between 3.0 and 4.5 adults/km^2. These specimens are now in the Department of Zoology, University College Dublin. The material from County Down is resident in the Ulster Museum (Belfast). The Gloucestershire material is held by the Ministry of Agriculture, Fisheries and Food and consisted of 25 animals of known age tagged as cubs. The Bristol material is currently resident in the Department of Zoology, University of Bristol, and consisted of 22 animals of known age tagged as cubs, and 22 animals estimated to be aged two years or over. The English populations are well documented (Cheeseman *et al.* 1988; Cresswell and Harris 1988). The Gloucestershire study site occupies 9.0km^2, where the population density is the highest reported in Europe (20.0 adults/km^2). A number of the Bristol sample came from a 5.5km^2 study site, with a density approximately half that of the Gloucestershire site, with the rest coming from in and around Bristol city (Harris 1984). The Scottish material is stored in the Royal Museum of Scotland (Edinburgh), and originated either from the Lothian or Speyside regions of Scotland.

The extent of genetic variation between badger populations was assessed by comparison of data from five sources as follows:

1. the polymorphism of 16 enzyme/protein systems in red cells and plasma of 170

badgers from Gloucestershire, as revealed by horizontal starch gel electrophoresis (Evans *et al.* 1989);

2. a report of 31 enzyme systems screened by horizontal starch gel electrophoresis, from a number of tissues of two badgers from Nickelsdorf, eastern Austria (Hartl *et al.* 1988);
3. a study of 15 enzyme systems by starch gel electrophoresis in muscle and liver tissue, in a sample of five badgers from Denmark (Simonsen 1982);
4. a study (J. Whitebrook and S. Harris, unpubl.) of eleven enzyme systems in blood of 75 badgers from Bristol, analysed by horizontal starch gel electrophoresis according to routine methods (Scopes 1968; Scandalios 1969; Harris and Hopkinson 1976; Roose and Gottlieb 1976; Brown *et al.* 1978);
5. a study of eleven enzyme systems in liver of 26 badgers from County Offaly, and five badgers from County Cork (Ireland) by vertical slab polyacrylamide gel electrophoresis (PAGE) (O'Corry-Crowe 1992).

Table 1. Number of crania used in the current study.

Origin	Male	Female	Unsexed	Total
Bristol	26	18	0	44
Gloucestershire	15	9	1	25
Offaly	10	9	1	20
Down	4	5	1	10
Scotland	0	0	23	23

Craniometric measurements

The following nine measurements were taken according to Wiig (1986): condylobasal length, basilar length, palatal length, maxillary tooth row length, mastoid breadth, zygomatic breadth, postorbital breadth, width of the postorbital constriction, and rostral breadth. Two additional measurements were also taken: minimum interorbital breadth and maximum breadth across the molars. Measurements were taken with dial callipers to the accuracy of 0.05mm. Suture closure (in particular the nasal, fronto-parietal, zygomatic, squamoso-parietal and pre-basisphenoid-basioccipital sutures), development of the saggital crest and relative breadth of the postorbital region could be used to separate the material into cubs, yearlings and adults. Only animals estimated (or known, in the case of the English material) to be aged two years or over were used in the analysis. This is the age at which growth of the eleven parameters measured has stabilised and it is impossible to discriminate year classes on the basis of these measurements alone using multivariate techniques (Il-Fituri 1992).

All the Irish and Scottish material was measured by one of the authors (JML), whilst the English material was measured by another (TJH). Examination of a sample of Irish animals (n = 25) measured by both of the above revealed no difference in the manner in which these measurements were taken that would influence the following analyses.

Owing to the high proportion of unsexed specimens (18.6% of total) the sexes were pooled. A chi-squared analysis showed no significant difference between the composition of the groups as regards proportion of known males and females. Similarly, it was observed that the degree of sexual dimorphism was similar in all populations utilised for this analysis. Analysis of sexual dimorphism within these populations, via linear discriminant function analysis, revealed that, at best, only 30% of specimens could be accurately sexed, thus highlighting the fact that differences in the composition of the samples (with respect to sex) were unlikely to be a major cause of variation between sites.

Statistical analysis

Data were screened for multivariate outliers using standard procedures (see Lynch 1993). Log-transformed data were then entered into a Canonical Variate Analysis (CVA) to assess variation between samples (in this case, sites of origin). Details of the methodology and usage of CVA are given in Marcus (1990) and Lynch (1993). Following the advice of Reyment (1990), no attempt was made to interpret the canonical axes obtained. Specimens were allocated to sites based on the discriminatory rules thus derived. The results of the reallocation process were compared to those expected by random sorting, taking account of unequal sample size, using Cohen's Kappa (Titus *et al.* 1984). Cohen's Kappa is an adjusted proportion of animals correctly reassigned to their group of origin and has an associated 95% confidence interval. All statistical procedures were carried out using the SAS™ package (version 6.06). The CVA was carried out using the DISCRIM procedure and a number of routines written in the SAS/IML language. The calculation of Cohen's Kappa was carried out using a SAS/IML program.

Statistical analysis of electrophoretic data was primarily carried out using the BIOSYS-1, Fortran-77 computer program (Swofford and Selander 1989). A locus was adjudged to be polymorphic if the proportion of the most abundant allele did not exceed 0.95. The proportion of polymorphic loci (P) was calculated for each population. The mean number of alleles per locus (A) was calculated and the average heterozygosity (H) (Nei 1978), expressed as the mean frequency of heterozygotes per locus, was estimated.

Comparison between populations requires data on not less than 14 homologous loci, studied by the same technique, rather than a small number of loci studied in a large number of individuals (Nevo 1978). This is necessary to avoid the effects of unequal evolutionary rates between proteins (Nei and Roychoudhury 1974; Nei 1978; Hartl *et al.* 1990) or differences in resolving power of different separation techniques (Ferguson 1980). Using a minimum of 14 loci to estimate the genetic heterogeneity (H and P) of a population, pairwise comparisons were made between Bristol and Gloucestershire, Bristol and Nickelsdorf, and Offaly and Cork to estimate genetic distance (D) and identity (I) (Nei 1978). These indices of genetic differences between populations are based on the accumulated number of gene substitutions per locus over all loci examined.

Since the minimum distance between the Bristol and Gloucestershire sites is 32km, the hypothesis that both samples were drawn from the same population was tested by subjecting the allele frequencies to the contingency chi-squared analysis of Workman and Niswander (1970). The same analysis was carried out on the Offaly and Cork samples, although they are derived from populations 190km apart.

RESULTS

Comparison of Irish, English and Scottish material

The Canonical Variate Analysis detected a significant separation (Wilks' Lambda = 0.439, $p < 0.0001$) between the three groups. The Mahalanobis distances (in standard deviation units) between groups are given in Table 2. All Mahalanobis distances were significant at $p < 0.0001$. Overall, this procedure resulted in, on average, 57.3% (95% CI = 11.8%) of animals being correctly reassigned to their site of origin.

Comparison of sites within countries

This analysis considers four groups: Offaly, Down, Bristol and Gloucestershire. The Canonical Variate Analysis resulted in three variates which aided population separation. The first ($p < 0.0001$) separated the Irish populations from the English, the second ($p < 0.0001$) Down from the rest, and the third Gloucestershire from the other three groups ($p < 0.02$). Thus, compared to the separation between the Irish sites, the difference between the two English samples is smaller. This pattern can be clearly seen in the Mahalanobis distances between populations (Table 3). All D^2 values were statistically significant ($p < 0.0001$). The overall separation was significant ($p < 0.0001$, Wilks' Lambda = 0.276, Fig. 1 and Fig. 2), and 36.5% (95% CI = 11.0%) of specimens were correctly reallocated to their site of origin.

Table 2. Morphological distances between sites: figures are D^2 (Mahalanobis distance in standard deviation units).

	ENGLAND	SCOTLAND
IRELAND	3.82	4.34
ENGLAND	-	1.71

Table 3. Morphological distances between sites: figures are D^2 (Mahalanobis distance in standard deviation units).

	OFFALY	BRISTOL	GLOUCS.
DOWN	4.59	9.85	6.77
OFFALY	-	5.67	3.42
BRISTOL		-	2.71

Analysis of the Irish and English sites in isolation (i.e. pairwise comparisons) reveals a similar pattern. Down badgers were significantly different from those from Offaly (Wilks' Lambda = 0.472, $p < 0.0002$, $D^2 = 7.19$, Kappa = 83.5%). Similarly, Gloucestershire badgers were significantly different from those from Bristol (Wilks' Lambda = 0.630, $p < 0.009$, $D^2 = 2.76$, Kappa = 60.2%). This strengthens the observation that the two Irish populations were relatively separate compared to those from England.

While much work remains to be done in this area, it is obvious that morphological differences exist between badger populations, and that, in general, such differences are related to geographic distance.

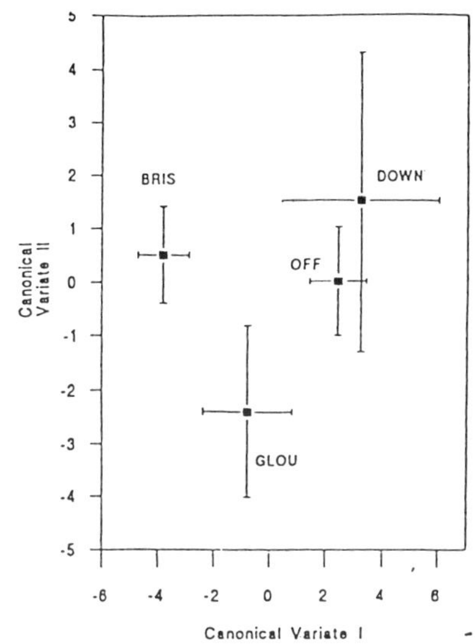

Fig. 1. Separation of badgers from Bristol, Offaly, Down and Gloucestershire on the basis of the first and second canonical variates extracted from eleven cranial measurements. Boundaries represent 95% confidence regions for the group centroid.

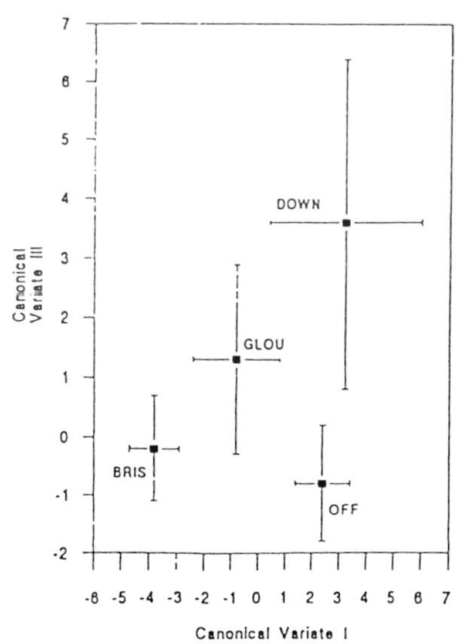

Fig. 2. Separation of badgers from Bristol, Offaly, Down and Gloucestershire on the basis of the first and third canonical variates extracted from eleven cranial measurements. Boundaries represent 95% confidence regions for the group centroid.

Genetic variation

The overall genetic variation in the six populations is summarised in Table 4. Of the 14 loci successfully resolved in the Offaly population, five (35.7%) were polymorphic by the 95% criterion for polymorphism. Three loci (21.4%) were polymorphic by the same criterion in the Cork population for the same set of (homologous) loci. In the sample from Bristol, the eleven enzyme systems represent 20 presumptive structural loci, of which five (25%) were found to be polymorphic. In the Gloucestershire sample, the 16 enzyme/protein systems represent 23 loci, one of which (4.3%) was polymorphic. Of the 31 enzyme systems, representing 35 successfully resolved loci, examined in the Nickelsdorf sample, one locus (2.8%) was polymorphic, while none of the 21 loci examined in the Danish study exhibited polymorphism. Average heterozygosity (H) for each population ranged from 0.000, in the Danish sample, to 0.132, in the Offaly sample (Table 4). The populations tended to fall into two categories: (i) those exhibiting little or no genetic variation (this includes the Danish, Austrian and Gloucestershire populations), and (ii) those exhibiting high genetic variability, which includes the Bristol and the two Irish populations.

The 52 loci examined in the six populations are listed in Table 5. Despite this range of loci, only two (Ldh-1 and Ldh-2) were examined in all six populations and, apart from Offaly, both were monomorphic. Excluding Denmark, five homologous loci were studied in the remaining populations. Between five and ten loci were studied in common for the various 3-way population comparisons. Pairwise comparisons ranged from 5 to 14 homologous loci in both members of the pair. But since at least 14 loci are required for a valid comparison, only three pairs could be examined. These were: Bristol vs Gloucestershire, Offaly vs Cork, and Bristol vs Austria.

Table 4. Summary of genetic variability in a number of badger populations.

Location	**N_i**	**N_l**	**P**	**H**	**A**
1. Offaly	26	14	0.357	0.132	1.43
2. Cork	5	14	0.214	0.103	1.21
3. Bristol	75	20	0.250	0.082	1.30
4. Gloucs.	170	23	0.043	0.020	1.09
5. Austria	2	35	0.029	0.014	1.03
6. Denmark	5	21	0.000	0.000	1.00

Note: N_i: number of individuals; N_l: number of loci; P: proportion of polymorphic loci; H: average heterozygosity; A: average number of alleles per locus. Sources of data: 1,2: this study, O'Corry-Crowe 1992; 3: Whitebrook and Harris, unpubl.; 4: Evans *et al.* 1989; 5: Hartl *et al.* 1988; 6: Simonsen 1982.

The gene pools of the Bristol and Gloucestershire populations are significantly different, as revealed by chi-squared analysis ($p < 0.001$). Contingency chi-squared analysis of allele frequencies between the two Irish populations also revealed significant separation of the gene pools ($p < 0.01$). The genetic distance (D) between Bristol and Gloucestershire populations, based on 14 homologous loci, was 0.036. The genetic distance (14 loci) between Offaly and Cork was also estimated at 0.036. The genetic distance between Bristol and Austria, generated from a similar number of loci, but not completely homologous to either of the above comparisons, was 0.085. As these three distance measurements are not directly comparable (see above), a dendrogram, representing genetic distances between groups of three or more populations cannot be constructed. Nevertheless, it is now clear that genetic differences between badger populations can be detected.

Table 5. List of enzyme systems investigated in the various studies. The number of loci successfully stained is given in parentheses, and a dash (-) indicates unsuccessful resolution or an inadequate level of staining.

Offaly: LDH(2), SOD(2), PGI(1), ADA(1), GOT(1), EST(2), MPI(-), MDH(1), AK(2), PEP A(1), PEP C(1)

Cork: LDH(2), SOD(2), PGI(1), ADA(1), GOT(1), EST(2), MPI(-), MDH (1), AK(2), PEP A(1), PEP C(1)

Bristol: LDH(2), SOD(2), PGI(2), EST(5), ME(2), ICD(1), 6PGD(1), GDH(1), G6PD(1), GLUDH(1), PGM(2)

Gloucestershire: LDH(2), SOD(2), PGI(1), ADA(-), GOT(-), EST(4), MPI(-), PEPA(1), PEP B(1), ME(-), ICD(1), 6PGD(1), GPD(1), PGM(2), HB(1), PT(4)

Austria: LDH(2), ADH(-), SOD(2), PGI(1), ADA(-), GOT(2), EST(2), MPI(1), MDH(2), AK(2), PEP 1(-), PEP 2(-), ME(2), ICD(2), 6PGD(1), GDH(1), G6P(-), GPD(1), GAPDH(1), XDH(1), GLUDH(1), PGM(2), SDH(1), DIA(-), CE(-), NP(1), HK(3), CK(2), ACP(1), ACY 1(1), FH(1), ACO(2)

Denmark: LDH(2), ADA(1), GOT(2), EST(2), MDH(2), AK(1), PEP(1), ICD(2), 6PGD(1), G6PD(1), DIA(2), ACP(1), CAT(1), PGK(1), GPT(1)

DISCUSSION

This study clearly shows that badger populations across Europe and the British Isles are morphologically and genetically distinct. There exists some debate as to how the genome and phenotype are coupled. Although osteometric traits have a high narrow-sense heritability (Atchley *et al.* 1981), body size has a strong non-genetic component. Multivariate morphometric techniques have in the past detected morphological differences between electrophoretically distinct population units for a number of mammalian species (e.g. *Thomomys bottae*; Patton and Smith 1990). Similarly, chromosomal 'races' of shrews (*Sorex araneus*; Searle and Thorpe 1987) and kangaroo rats (*Dipodomys agilis*; Best *et al.* 1986) are morphologically distinct. On the other hand, a number of researchers quite rightly believe that morphological similarity cannot be taken to indicate genomic similarity (e.g. Baverstock and Adams 1987).

Whilst a number of factors may contribute to the patterns of morphological variation between populations, it remains difficult to separate their effects. Non-representative data might be generated by both inter-observer variation and the sampling method, particularly if there are temporal fluctuations in, ontogenetic changes in, and pathological sources of variation within these study populations. It is argued that the effects of these factors have been minimised. Firstly, only skulls from adults have been considered. Skull growth is essentially complete by the third year of life and multivariate analysis has failed to detect an age-related component in the pattern of variation in adults (Il-Fituri 1992). Secondly, the samples of skulls representing each population contain adults of different ages collected in different years. Thus sampling has not been restricted to a single cohort. This negates the effect of morphometric variation which might be attributable to variation in early growth rate owing to suboptimal nutrition. Thirdly, although the effects of disease or parasite burden on cranial size and shape are unknown, infestation of some mustelids by the nematode *Skrjabingylus nasicola* results in deformed crania (King and Moody 1982). No evidence of such damage was found in the crania used in this study. Furthermore, given the ecological characteristics of the badger, it is unlikely that the differences between the populations are related to prey type or size. It is thus concluded that interpopulation differences described here are primarily an indicator of genetic differences between populations.

We believe that morphometric data of this type can help answer biogeographic questions such as the past history of the Irish carnivoran fauna. However, there are at present insufficient data to address this problem and indeed that of microgeographic variation (but see Lynch 1993; Lynch and Hayden, in press).

While it is relatively straightforward, when appropriate samples and methods are available, to compare morphology across populations (and such data can easily be collected from studies designed for another purpose), the same cannot be said for the study of enzyme polymorphisms. To compare genetic variation several *a priori* decisions must be taken with respect to tissue type to be collected, loci to be examined and separation method to be utilised. It is clear from the studies reviewed here (both published and unpublished) that a comparison between populations was never the primary intention (as indicated by the fact that only 2 of 51 loci were common to all studies). Nevertheless, badger populations are obviously genetically distinct, and in the cases under review demonstrate mean heterozygosities ranging from 0.000 to 0.132, typical of 'low' and 'high' genetic variability (Nevo 1978; Baccus *et al.* 1983; Testa 1986).

It should be noted that, while five populations were available for morphological study and six populations were utilised in the genetic analysis, badgers from only three areas were common to both studies. This, along with the lack of comparability between the genetic studies, prevented us from examining the congruence between genetic and morphologic divergence. We can, however, state that morphological distance is, to some extent, related to geographical separation, as is genetic distance, though genetic and morphologic distances cannot be directly compared. It is hardly surprising, therefore, that the largest genetic distances were between the British and Austrian populations when one considers the magnitude of the geographic distance separating these populations and the separation of Britain from the continental landmass for at least the last 9,000 years (Yalden 1990).

Mainland badgers in Britain and Ireland constitute two distinct geographic populations. However, the social organisation and limited dispersal ability of this species (Kruuk 1978; Cheeseman *et al.* 1988) restricts the movement of individuals to a new area, thus promoting local genetic differentiation. This isolation by distance, with concomitant restriction in gene flow, may explain, in part, the genetic differences between Offaly and Cork, and between the two English populations. The Bristol badgers may be a relict population enclosed by post-First World War development (Harris 1984).

The extent of gene flow between this and the surrounding rural population may therefore be limited. Furthermore, local extinctions with subsequent recolonisation may promote genetic differences between sites owing to 'founder effects', in addition to genetic drift from partial reproductive isolation. This does not exclude the possibility that adaptation to local conditions contributes to genetic differentiation. The significant difference in gene pools between Bristol and Gloucestershire may reflect a response to different selective pressures from environmental or demographic factors in town and country.

What are the implications of the observed genetic differences between populations? Firstly, it is now clear that a more detailed examination of variation within the British Isles can be expected to contribute to an understanding of the origin of the Irish fauna. Secondly, they introduce the possibility that another element is involved in the aetiology of tuberculosis in badgers. There is evidence of considerable variation in the prevalence of *Mycobacterium bovis* infection in badgers in different counties in Britain, despite similar badger densities (Cheeseman *et al.* 1989a), and the prevalence of disease in badgers from some parts of Ireland (this volume) indicates a generally high level of disease compared to Britain. This raises the possibility of genetic differences in resistance to tuberculosis infection in different badger populations across Britain and Ireland (O'Corry-Crowe 1992).

Resistance to tuberculosis has been studied in inbred strains of rabbits (Lurie 1964; Lurie and Dannenberg 1965) and mice (reviewed by Skamene 1989). In mice, resistance to tuberculosis is controlled by a single dominant autosomal gene, called the *Bcg* gene, which acts by controlling T-cell independent macrophage activation (Skamene 1989). Evidence suggests that similar genetic control of resistance occurs in rabbits (Lurie 1964; Lurie and Dannenberg 1965) and in humans (Skamene 1989). A mechanism such as this could explain some of the variation in prevalence of tuberculosis in badgers from different areas/countries. Some caution is needed in the interpretation of this information. Methods of screening for tuberculosis in badgers in Britain and Ireland are not directly comparable. Since inspection of carcasses is the method routinely used to examine for the presence of infection in Irish badgers (Fagan 1993), the difference in prevalence between the two countries is probably greater than that observed. A further difficulty is that the prevalence revealed in samples taken at a particular time will mask the dynamic pattern of the disease in badgers (Cheeseman *et al.* 1989b). While there are many other factors which may play a role in the presentation of the disease in the badger population (Cheeseman *et al.* 1989a), it is possible that bovine TB might select for alleles conferring increased resistance within badger populations.

Now that the existence of population-specific patterns of craniometric variation has been established it is appropriate to extend the biogeographical study by sampling further populations. While it must be noted that the factors responsible for the craniometric variation are not known, it is likely that there is an important genetic component. Similarly the relative importance of the factors responsible for the genetic differences between the populations considered here is also unknown, but comparative analyses are now becoming possible (Evans *et al.* 1989; O'Corry Crowe 1992; J. Whitebrook and S. Harris, unpubl.). The extent and the significance of genetic variation within and between populations are, as yet, insufficiently known but may be of importance in relation to regional differences in the extent of bovine tuberculosis in badger populations.

ACKNOWLEDGEMENTS

We would like to thank the following for their help in the completion of this work: Mr T. Duffy and Dr L. Dolan (ERAD), Mr J. Eves, Mr J. McCarthy and Mr M. Hennessy (the Department of Agriculture and Food), the staff of the Regional Veterinary Laboratories, Ms R. Whelan and Ms A. Il-Fituri (UCD), Dr W. Cresswell (Bristol), Dr A. Kitchener (Royal Museum of Scotland), and Mr T. Bruton (Ulster Museum). We are grateful to Dr A. Nolan (Central Veterinary Laboratories, Weybridge) and Dr G. Smith (MAFF) for comments on an earlier draft of this paper. Dr L. Marcus (American Museum of Natural History) provided a number of the routines used to analyse the data. We further acknowledge funding from EOLAS to JML to defray travel expenses. The material from Bristol was collected whilst SH was in receipt of grants from the Ministry of Agriculture, Fisheries and Food, the Natural Environment Research Council and the Joint Nature Conservation Committee.

REFERENCES

ATCHLEY, W. R., RUTLEDGE, J. J. and COWLEY, D. E. 1981 Genetic components of size and shape. II. Multivariate covariance patterns in the rat and mouse skull. *Evolution* **35**, 1037-55.

BACCUS, R., RYMAN, N., SMITH, M. H., REUTERWALL, C. and CAMERON, D. 1983 Genetic variability and differentiation in large grazing mammals. *Journal of Mammalogy* **64**, 109-20.

BAVERSTOCK, P. R. and ADAMS, M. 1987 Comparative rates of molecular, chromosomal and morphological evolution in some Australian vertebrates. In K.S.W. Campbell and M. F. Day (eds), *Rates of evolution,* 175-88. London. Allen and Unwin.

BEST, T. L., O'SULLIVAN, R. M., COOK, J. A. and YATES, T. L. 1986 Chromosomal, genic and morphologic variation in the Agile Kangaroo Rat, *Dipodomys agilis* (Rodentia: Heteromyidae). *Systematic Zoology* **35**, 311-24.

BROWN, A. H. D., NEVO, E., ZOHARY, D. and DANGAN, O. 1978 Genetic variation in natural populations of wild barley (*Hordeum spontaneum*). *Genetica* **59**, 97-108.

CHEESEMAN, C. L., CRESSWELL, W. J., HARRIS, S. and MALLINSON, P. J. 1988 Comparison of dispersal and other movements in two badger (*Meles meles*) populations. *Mammal Review* **18**, 51-9.

CHEESEMAN, C. L., WILESMITH, J. W. and STUART, F. A. 1989a Tuberculosis: the disease and its epidemiology in the badger, a review. *Epidemiological Information* **103**, 113-25.

CHEESEMAN, C. L., WILESMITH, J. W., STUART, F. A. and MALLINSON, P. J. 1989b Dynamics of tuberculosis in a naturally infected badger population. *Mammal Review* **18**, 61-72.

CRESSWELL, W. J. and HARRIS, S. 1988 Foraging behaviour and home-range utilization in a suburban badger (*Meles meles*) population. *Mammal Review* **18,** 37-49.

EVES, J. 1993 The East Offaly Badger Research Project: an interim report. In T. J. Hayden (ed.), *The badger,* 166-73. Dublin. Royal Irish Academy.

EVANS, P. G. H., MACDONALD, D. W. and CHEESEMAN, C. L. 1989 Social structure of the Eurasian badger (*Meles meles*): genetic evidence. *Journal of Zoology* **218**, 587-95.

FAGAN, J. 1993 Tuberculosis in badgers in Ireland: pathology. In T. J. Hayden (ed.), *The badger,* 117-22. Dublin. Royal Irish Academy.

FERGUSON, A. 1980 *Biochemical systematics and evolution.* New York. Halsted Press.

HARRIS, H. and HOPKINSON, D. A. 1976 *Handbook of enzyme electrophoresis in human genetics.* Amsterdam. North Holland Publishing Company.

HARRIS, S. 1984 Ecology of urban badgers (*Meles meles*): distribution in Britain and habitat selection, persecution, food and damage in the city of Bristol. *Biological Conservation* **28**, 349-75.

HARTL, G. B., WELLING, R., GRILLITSCH, M. and KLAUSEK, E. 1988 Biochemical variation in Mustelidae: are carnivores genetically less variable than other mammals? *Zoologischer Anzeiger* **221**, 81-90.

HARTL, G. B., WELLING, R. and SUCHAUTRUNK, F. 1990 On the biochemical systematics of selected mammalian taxa: empirical comparison of qualitative and quantitative approaches in the evaluation of protein electrophoretic data. *Zeitschrift für Zoologische Systematik und Evolutionsforschung* **28,** 191-216.

HELL, P. and PAULE, L. 1989 Craniometrical investigation of the European badger (*Meles meles*) from the Slovak Carpathians, Czechoslovakia. *Folia Zoologica* **38**, 307-23.

HUSON, L. W. and PAGE, R. J. C. 1980 Multivariate geographical variation of the red fox (*Vulpes vulpes*) in Wales. *Journal of Zoology, London* **191**, 453-9.

IL-FITURI, A. 1992 An evaluation of indicators of age in Irish badgers. Unpublished M.Sc. thesis, National University of Ireland.

KING, C. M. and MOODY, J. E. 1982 The biology of the stoat (*Mustela erminea*) in the National Parks of New Zealand. VI. Infestation with *Skrjabingylus nasicola. New Zealand Journal of Zoology* **9**, 131-40.

KRUUK, H. 1978 Spatial organisation and territorial behaviour of the European badger (*Meles meles*). *Journal of Zoology, London* **184**, 1-19.

LURIE, M. B. 1964 *Resistance to tuberculosis: experimental studies in natural and acquired defensive mechanisms.* Harvard University Press.

LURIE, M. B. and DANNENBERG, A. M. 1965 Macrophage function in infectious disease with inbred rabbits. *Bacteriological Review* **29**, 466-76.

LYNCH, J. M. 1993 Craniometric variation in selected Mustelidae. Unpublished Ph.D. thesis, National University of Ireland.

LYNCH, J. M. and HAYDEN, T. J. (in press) Multivariate morphometrics and the biogeography of Irish mustelids. In M. J. Costello (ed.), *Biogeography of Ireland: past, present, and future.* Dublin. Irish Biogeographical Society.

MARCUS, L. 1990 Traditional morphometrics. In F.J. Rholf and F.L. Bookstein (eds), *Proceedings of the Michigan morphometrics workshop,* 77-122. Ann Arbor. University of Michigan, Museum of Zoology.

MANNING, T. H. 1971 Geographical variation in the polar bear *Ursus maritimus* Phipps. *Canadian Wildlife Service Report Series* **13**, 1-27.

NEAL, E. G. and CHEESEMAN, C. L. 1991 Badger (*Meles meles*). In G. B. Corbet and S. Harris (eds), *The handbook of British mammals* (3rd edn), 415-23. Oxford. Blackwell Scientific Publications.

NEI, M. 1978 Estimation of average heterozygosity and genetic distance from a small number of individuals. *Genetics* **89**, 583-90.

NEI, M. and ROYCHOUDHURY, A. K. 1974 Sampling variances of heterozygosity and genetic distance. *Genetics* **76**, 379-90.

NEVO, E. 1978 Genetic variation in natural populations: patterns and theory. *Theoretical Population Biology* **13**, 121-74.

O'CORRY-CROWE, G. 1992 Social and genetic structure of a population of badgers (*Meles meles* L.) in central Ireland. Unpublished Ph.D. thesis, National University of Ireland.

PATTON, J. L. and SMITH, M. F. 1990 The evolutionary dynamics of the pocket gopher *Thomomys bottae*, with emphasis on Californian populations. *University of California Publications in Zoology* **123**, 1-161.

REYMENT, R. A. 1990 Reification of classical multivariate analysis in morphometry. In F. J. Rholf and F. L. Bookstein (eds), *Proceedings of the Michigan morphometrics Workshop,* 123-44. Ann Arbor. University of Michigan, Museum of Zoology.

ROOSE, M. L. and GOTTLIEB, L. D. 1976 Genetic and biochemical consequences of polyploidy in *Tragopogon. Evolution* **30**, 818-30.

SCANDALIOS, J. C. 1969 Genetic control of multiple molecular forms of enzymes in plants: a review. *Biochemical Genetics* **3**, 37-79.

SCOPES, R. K. 1968 Methods for starch-gel electrophoresis of sarcoplasmic proteins. *Biochemical Journal* **107**, 139-50.

SEARLE, J. B. and THORPE, R. S. 1987 Morphometric variation of the common shrew (*Sorex araneus*) in Britain, in relation to karyotype and geography. *Journal of Zoology, London* **212**, 373-7.

SKAMENE, E. 1989 *Review of infectious diseases,* volume II (supplement 2), 5394-9.

SIMONSEN, V. 1982 Electrophoretic variation in large mammals. II. The red fox (*Vulpes vulpes*), the stoat (*Mustela erminea*), the weasel (*Mustela nivalis*), the polecat (*Mustela putorius*), the pine marten (*Martes martes*), the beech marten (*Martes foina*), and the badger (*Meles meles*). *Hereditas* **96**, 299-305.

SWOFFORD, D. L. and SELANDER, R. B. 1989 *BIOSYS-1: A computer program for the analysis of allelic variation in population genetics and biochemical systematics.* Illinois. Illinois Natural History Survey.

TESTA, J. W. 1986 Electromorph variation in Weddel seals (*Leptonychotes weddelli*). *Journal of Mammalogy* **67**, 606-10.

TITUS, K., MOSHER, J. A and WILLIAMS, B. K. 1984 Chance-corrected classification for use in discriminant analysis: ecological applications. *American Midland Naturalist* **111**,1-8.

WIIG, Ø. 1986 Sexual dimorphism in the skull of minks (*Mustela vison*), badgers (*Meles meles*) and otters (*Lutra lutra*). *Zoological Journal of the Linnean Society* **87**, 163-79.

WORKMAN, P. L. and NISWANDER, J. D. 1970 Population genetics of Southwestern Indian tribes. II. Local genetic differentiation in the Panpago. *American Journal of Human Genetics* **22**, 24-49.

YALDEN, D. W. 1990 History of the fauna In G. B. Corbet and S. Harris (eds). *The handbook of British mammals* (3rd edn), 7-18. Oxford. Blackwell Scientific Publications.

In: Hayden, T.J. (ed.) 1993 *The badger*, 108-116. Dublin. Royal Irish Academy.

BADGERS AND BOVINE TUBERCULOSIS IN IRELAND: A REVIEW

L.A. Dolan

ERAD/Teagasc, Tuberculosis Investigation Unit, University College Dublin, Veterinary College, Ballsbridge, Dublin 4

ABSTRACT

A review is presented of the main conclusions of the principal reports concerning badgers and bovine tuberculosis in Great Britain and Ireland. In these reports it is widely accepted that badgers can transmit tuberculosis to cattle, but both the mode and extent of transmission remain matters for debate. The number of badgers examined and those which disclosed tuberculous lesions on gross post-mortem examination in Ireland are tabulated for each county from 1980 to 1989. The distribution of lesion type within the badgers is also recorded. Tuberculous badgers were recorded in every county in the country; therefore it is suggested that tuberculosis is endemic in the badger population. An effective vaccine is suggested as a possible solution to the problem of tuberculosis in badgers and cattle.

INTRODUCTION

Tuberculosis was first recorded in badgers in Switzerland (Bouvier *et al.* 1962). Tuberculous badgers were first identified in 1971 in the south-west of England (Muirhead *et al.* 1974) and in 1973 in west Cork in Ireland (Noonan *et al.* 1975). In Ireland the badger and its habitat became protected under the Wildlife Act of 1976 at the same time as similar legislation was enacted in Britain.

Since tuberculosis was first identified in badgers in the south-west of England, extensive programmes have been undertaken to investigate the relationship between infected badgers and infected cattle. This relationship has been the subject of two major reports commissioned by the government in Britain. The first was the Zuckerman report, published in 1980, some of the main conclusions of which were (1) that the infected badger was a secondary source of tuberculosis for cattle, (2) that the gassing of infected social groups should be resumed but that research should be carried out to improve its efficiency, (3) that the consultative panel drawn from interested organisations should remain and a liaison officer be appointed, (4) that research on badger dynamics should be continued, (5) that badger removal operations should be instigated more rapidly, and (6) that the situation should be reviewed after three years and the findings should be published.

The second report was the Dunnet Report, published in 1986, which concluded that the badger was the primary source of tuberculosis for cattle, and that there was no need to do further research on the transmission of tuberculosis from badgers to cattle, as this was proven. It was recommended, as an interim measure, that badgers should only be trapped on the breakdown farm or that part of the farm where the breakdown occurred unless the level of herd breakdowns went above a certain threshold. Further recommendations were that research should be continued on (1) the population dynamics of the badger, (2) an *in vivo* test in the badger and (3) badger vaccination. It also concluded that 40% of the improvement in tuberculosis levels in cattle could be attributed to the badger removal operations.

A report was also published by the Badger Wildlife Link Group in 1984, a group which comprised all of the Animal Welfare and Conservation bodies in Britain. They accepted that tuberculosis could be transmitted from badgers to cattle and that, 'in the absence of any other identified cause, if an infected badger or badgers were found within a radius of 1.5km of a herd breakdown then it was reasonable to conclude that the badgers were responsible for the herd breakdown'. The main areas of debate were the exact mode of transmission from badgers to cattle and the extent to which the infected badgers were the source of herd breakdowns. They were also concerned at the numbers of non-infected badgers being killed and consequently were very anxious that an *in vivo* test be developed.

Cheeseman *et al.* (1989) reviewed the data concerning badgers and tuberculosis in Britain and concluded (1) that tuberculosis was endemic in the badger population, (2) that the badger was an ideal maintenance host for tuberculosis, (3) that tuberculosis was self-sustaining in the badger population, and (4) that the disease was cyclic in the badger population.

The Economic and Social Research Institute (ESRI) published a report on badgers and bovine tuberculosis in Ireland (O'Connor and O'Malley 1989) which included a review of the situation in Britain and the results of field studies carried out in five counties in Ireland. The field studies all recorded an immediate and substantial drop in disease levels in the cattle once infected badgers were removed. One of the conclusions of this report was that tuberculosis in cattle might not be eradicated from parts of Ireland unless an effective badger control policy was implemented.

Morris and Pfeiffer (1990) in their review of the National Tuberculosis Eradication Programme in Ireland concluded that tuberculous badgers, by seeding infection into herds throughout the country, were the underlying driving factor causing special difficulties in achieving eradication of tuberculosis from the cattle population. They also concluded that tuberculosis has probably been present in the badger population in Ireland for at least 30 years and hence has reached a stable geographical spread. They suggested that an effective vaccine would be required for badgers or cattle (preferably both) in order to facilitate eradication and that it would be difficult to see a complete solution without such a vaccine. To this end they strongly urged that the authorities in Ireland should support an international project to develop such a vaccine. In the interim they suggested annual cropping of badger numbers, targeted at the infected social groups, and the possibility of replacing tuberculous badger populations with disease-free badgers, as a pilot study.

Downey (1990) in his review of the Bovine Tuberculosis Eradication Scheme concluded that at least part of the problem of continuing herd breakdowns in areas with a high prevalence of disease was due to infected badgers. Further, he considered that if a more effective and efficient means of removal of infected badgers is not permitted (other than snaring), then the targeted reduction in the levels of tuberculosis in cattle may not be achieved.

MATERIALS AND METHODS

Selective survey of badgers in Ireland (1980-9)

The survey that is reported in this paper is comprised of data on badgers which were subjected to a post-mortem examination during the period 1980 to 1989 and were correlated with an individual farm or a defined area in each county.

The badgers examined from 1980 to 1984 were either road casualties or were found dead on farms. From 1984 onwards, licences to capture badgers were obtained from the Wildlife Service for the purpose of investigating the possible involvement of tuberculous badgers in herd breakdowns. Badgers were snared in areas with high numbers of infected herds which had not responded to standard procedures such as repeated test and slaughter of infected cattle, disinfection of premises etc., or where sudden and extensive unexplained outbreaks occurred. Therefore, the sample of badgers which was examined was selective and not random.

The badgers were snared and killed by shooting. The carcass was tagged and submitted to a regional veterinary laboratory for post-mortem examination (p.m.e.). In most cases the badger carcass was examined for evidence of gross visible abnormalities only and, if any were seen, specimens from the affected tissues were subjected to histopathological and/or cultural examination for evidence of tuberculosis. In some cases pooled samples of lymph nodes were submitted for histological and cultural examinations from badgers in which there were no visible lesions. All of the badgers captured in County Cavan, however, were subjected to a full, detailed p.m.e. The result of the p.m.e. was correlated with the appropriate farm identification number, i.e. the herd number, or with the area number, i.e. district electoral division number (DED), if captured on other than farmland.

A considerable number of the badger carcasses or selected specimens were assembled at the regional veterinary laboratories for further study at the Zoology Department, University College, Dublin. Blood samples were also collected from badgers for use in the development of a blood-based test.

RESULTS

The total number of badgers examined and which correlated with a herd number or DED number from 1980 to 1989 was 3,909 (Table 1). Of these, 664 (17%) showed evidence of infection with tuberculosis. Badgers snared in the East Offaly Research Project which commenced in County Offaly in 1989 are not included in these national figures. The distribution of the principal lesion sites is recorded in Fig. 1.

Table 1. The number of badgers examined for evidence of tuberculosis, by year, in a selective study in the Republic of Ireland (1980-9) (mainly gross post-mortem examination).

Year	Negative	Positive(%)	Total
1980	2	0	2
1981	12	4 (25)	16
1982	6	1 (14)	7
1983	9	2 (18)	11
1984	46	17 (27)	63
1985	226	51 (18)	277
1986	424	117 (22)	541
1987	727	127 (15)	854
1988	1299	225 (15)	1524
1989	494	20 (20)	614
Total	3245	664 (17)	3909

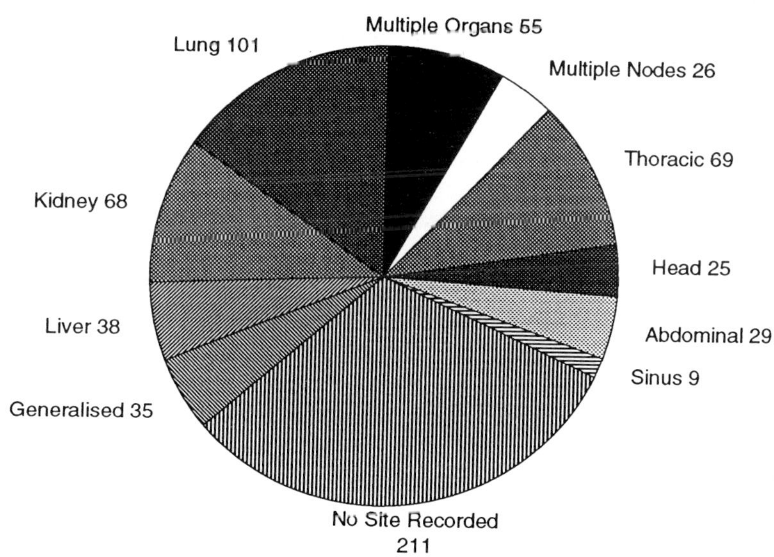

Fig. 1. Site of tuberculous lesion/s in 664 badgers at post-mortem 1980-9.

The lesion sites were classified in the order which suggested the greatest risk of shedding of bacteria, i.e. generalised, sinus, multiple organs etc. Where an organ and its associated lymph node were both affected, the site recorded was that of the organ. No lesion site was recorded for 211 of the 664 infected badgers. The lungs and kidneys were the most commonly affected organs and the thoracic lymph nodes were the most common lesion sites in the lymph system. The distribution of the badgers within counties is recorded in Table 2. Table 3 presents the number of badgers, by county, for which only a DED number could be ascribed and which were mostly road casualties or were found dead. The overall lesion rate in this group was 15.5%, with a wide variation between counties (0-55%).

Table 2. The total number of badgers examined together with their post-mortem examination (mainly gross post-mortem examination) results for each county in the Republic of Ireland (1980-9).

County	No. examined	No. positive	% positive
Carlow	42	12	29
Cavan	133	51	38
Clare	182	28	15
Cork	234	45	19
Donegal	65	9	14
Dublin	9	1	11
Galway	257	46	18
Kerry	173	32	19
Kildare	41	10	24
Kilkenny	74	14	19
Laois	22	2	9
Leitrim	66	7	11
Limerick	211	19	9
Longford	589	89	15
Louth	29	7	24
Mayo	31	7	23
Meath	28	9	32
Monaghan	28	13	46
Offaly	700	134	19
Roscommon	308	45	15
Sligo	87	8	9
Tipperary	274	31	11
Waterford	4	2	50
Westmeath	273	35	13
Wexford	20	2	10
Wicklow	29	6	37
Totals	3909	664	17

Table 3. The number of badgers (414) examined out of a total of 3909, by county, for which no individual herd number was available, i.e. road casualties, found dead etc., together with the post-mortem examination result (1980-9).

County	No. examined	No. positive	% positive
Carlow	4	0	0
Cavan	11	6	55
Clare	22	7	32
Cork	23	8	35
Donegal	5	0	0
Dublin	7	0	0
Galway	2	0	0
Kerry	24	5	21
Kildare	4	1	25
Kilkenny	18	7	39
Laois	0	0	0
Leitrim	8	0	0
Limerick	54	2	4
Longford	43	3	7
Louth	4	1	25
Mayo	4	2	50
Meath	19	6	46
Monaghan	0	0	0
Offaly	21	1	5
Roscommon	22	3	14
Sligo	24	1	4
Tipperary	31	3	14
Waterford	0	0	0
Westmeath	49	6	12
Wexford	2	0	0
Wicklow	13	2	33
Totals	414	64	15.5

A majority of these badgers were likely to have been submitted for examination from areas in which some snaring had taken place or from areas where an individual or individuals had taken a particular interest in the possible involvement of badgers in herd breakdowns. It may not, therefore, be a totally representative sample.

DISCUSSION

In the major reports published in Britain and Ireland concerning badgers and bovine tuberculosis it is accepted that tuberculous badgers can transmit infection to cattle (Zuckerman 1980; Badger Wildlife Link Group 1984; Dunnet *et al.* 1986; O'Connor and O'Malley 1989). This has already been demonstrated experimentally in Britain (Little *et al.* 1982). The exact mode of transmission has not been established, but a strong correlation between terminally ill tuberculous badgers and sudden and severe herd breakdowns has been a common feature of the continuing problem of bovine tuberculosis in particular areas in Ireland. A comparable situation has been reported from New Zealand, where tuberculous possums have been acknowledged as the cause of the persistence of bovine tuberculosis (MAF 1986). In Great Britain it has been concluded that tuberculosis is endemic and self-sustaining in the badger population (Cheeseman *et al.* 1989). The infected badgers are found mostly in the south-west of England where tuberculosis has persisted in the cattle population. In Ireland, tuberculous badgers have been identified in every county since 1980 and therefore, it is reasonable to conclude that tuberculosis is also endemic in the Irish badger population. However, the infection is more widespread here than in Great Britain.

Morris and Pfeiffer (1990) concluded that tuberculosis was being seeded into the cattle herds in Ireland from the infected badger population, thereby causing special difficulties for the eradication of the disease in the cattle population. O'Connor and O'Malley (1989) also concluded that infected badgers were responsible for the persistence of tuberculosis in cattle herds in certain parts of Ireland. This was based on the reported fieldwork from five counties. However, many other counties in the country have similar levels of infection in their badger populations, and therefore it would seem logical that if similar projects were undertaken in these counties, the same conclusions would be reached.

While the mode of transmission of tuberculosis from badgers to cattle remains a matter for debate, it is known that tuberculous lesions in the badger contain high numbers of mycobacteria, are surrounded by a thin capsule and tend to coalesce in a progressive fashion, thereby forming extensive lesions which lead to death. Such lesions are commonly found in the lungs and kidneys (Gallagher *et al.* 1976; Little *et al.* 1982). Occasionally, superficial gland lesions rupture to the exterior and form weeping sinuses which discharge material heavily contaminated with *M. bovis*. Similarly, infected bite wounds discharge such infected material (Gallagher and Nelson 1979). It has also been recorded that as many as 300,000 *M. bovis* have been found in 1mL of urine and 200,000 in 1mL of sputum (MAFF 1979). This means that there is ample opportunity for tuberculous badgers to cause considerable contamination of pasture, feedstuffs or drinking water, and/or at the terminal stage of illness to cause infection in the associated cattle as a result of direct contact. The range of lesion type and site in the badger in this study was similar to that reported in Britain. Cattle, on the other hand, tend to localise *M. bovis* quite well in lesions which are commonly calcified, contain fewer organisms and are surrounded by thick fibrous capsules. In a majority of cases the infection progresses, but only slowly, delayed by an acquired cellular immunity (Jubb *et al.* 1985).

The fact that tuberculous badgers have been identified in all 26 counties in the Republic of Ireland since 1980 would seem to indicate that the disease has been endemic in the badger population for many years. The badgers are infected with *M. bovis* and it is probable that they became infected from the cattle population. Grange *et al.* 1990 found that the strains of *M. bovis* isolated from tuberculous cattle and badgers in Ireland were the same. They identified two different strains using the Dulwich typing scheme, but the same strain of *M. bovis* was isolated from badgers and cattle in the same areas of the country.

As tuberculosis is so widespread in the Irish badger population it would not be possible to remove all infected badger social groups, so the suggestion of Morris and Pfeiffer (1990) that an effective vaccine be developed would appear to be the best way forward in the long term. Research has been in progress to develop and deliver a vaccine to badgers in the field (Stanford *et al.* 1993; McCarthy 1993). In the short term, identified infected badger social groups will have to be removed where tuberculosis in the associated cattle population has failed to respond to conventional procedures.

The level of tuberculosis in casualty badgers, for which only a DED number was available, was very high (>15%). However, this would not have represented a totally random sample, nor did it have an adequate geographical spread. Consequently, in 1991, a nationwide operation to collect and examine all road-casualty badgers will be undertaken as part of the Bovine Tuberculosis Eradication programme.

ACKNOWLEDGEMENTS

My thanks are due to Professor J.D. Collins and fellow members of the Tuberculosis Investigation Unit for advice and constructive criticism, to J. Crilly for data assembly, to K. Lynch for computerised data and map, and to Ms S. McAuley for typing the paper. Especial thanks to all the staff at the district veterinary offices and the regional veterinary laboratories for their assistance and cooperation.

REFERENCES

BADGER WILDLIFE LINK WORKING GROUP 1984 *Badgers, cattle and bovine tuberculosis.* A report to the Minister of Agriculture's Bovine Tuberculosis Review Group, London.

BOUVIER, G., BURGISSER, H. and SCHNEIDER, P.A. 1962 Observations sur les maladies du giber, des oiseaux et des poissons. *Schweizer Archiv für Tierheilkunde* **104**, 440-50.

CHEESEMAN, C.H., WILESMITH, J.W. and STUART, F.A. 1989 Tuberculosis: the disease and its epidemiology in the badger: a review. *Epidemiological Information* **103**, 113-25.

DOWNEY, L. 1990 *Ireland's TB problem: what can and must be achieved.* Dublin. ICOS.

DUNNET, G.M., JONES, D.M. and MCINERNEY, J.P. 1986 *Badgers and bovine tuberculosis.* London. HMSO.

GALLAGHER, J. and NELSON, J. 1979 Causes of ill health and natural death in badgers in Gloucestershire. *Veterinary Record* **105,** 546-51.

GALLAGHER, J., MUIRHEAD, R.H. and BURN, K.J. 1976 Tuberculosis in wild badgers (*Meles meles*) in Gloucestershire: pathology. *Veterinary Record* **98**, 9-14.

GRANGE, J.M., COLLINS, J.D., O'REILLY, L.M., COSTELLO, E. and YATES, M.D. 1990 Identification and characteristics of *Mycobacterium bovis* isolated from cattle, badgers and deer in the Republic of Ireland. *Irish Veterinary Journal* **43**, 33-5.

JUBB, K.V.F., KENNEDY, P.C. and PALMER, N. 1985 *Pathology of domestic animals* (3rd Edn), **vol. 2,** 498-502. London. Academic Press Inc.

LITTLE, T.W.A., NAYLOR, P.F and WILESMITH, J.W. 1982 Laboratory studies of *M. bovis* infection in badgers and calves.*Veterinary Record* **111**, 550-7.

MAF 1986 Cattle TB. *Surveillance* **13,** Wellington, New Zealand. Ministry of Agriculture and Fisheries.

MAFF 1979 *Bovine tuberculosis in badgers: third report.* Ministry of Agriculture Fisheries and Food. London. HMSO.

MCCARTHY, J. 1993 The badger vaccination trial in west Cork: progress report. In T.J. Hayden (ed.), *The badger,* 181-8. Dublin. Royal Irish Academy.

MORRIS, R.S. and PFEIFFER, D. 1990 Report of Consultancy on Bovine Tuberculosis Eradication Programme in Ireland. Dublin. ERAD. Department of Agriculture and Food.

MUIRHEAD, R.H., GALLAGHER, J. and BURN, K.J. 1974 Tuberculosis in wild badgers in Gloucestershire: epidemiology. *Veterinary Record* **95**, 552-5.

NOONAN, N.L., SHEANE, W.D., HARPER, L.R. and RYAN, P.J. 1975 Wildlife as a possible reservoir of bovine TB. *Irish Veterinary Journal* **29**, 1.

O'CONNOR, R. and O'MALLEY, E. 1989 *Badgers and bovine tuberculosis.* Dublin. Economic and Social Research Institute/ERAD.

STANFORD, J., STAINSBY, K. and MAHMOOD, K.H. 1993 Could vaccination be the answer to the badger tuberculosis problem? In T.J. Hayden (ed.), *The badger,* 174-80. Dublin. Royal Irish Academy.

ZUCKERMAN, LORD 1980 *Badgers, cattle and tuberculosis.* London. HMSO.

In: Hayden, T.J. (ed.) 1993 *The badger*, 117-122. Dublin. Royal Irish Academy.

TUBERCULOSIS IN BADGERS IN IRELAND: PATHOLOGY

J. Fagan

Regional Veterinary Laboratory, Coosan, Athlone, Co. Westmeath

ABSTRACT

Routine diagnosis of tuberculosis (TB) in badgers in Ireland is carried out by necropsy (post-mortem examination) at the veterinary laboratories of the Department of Agriculture and Food. In a special survey of 83 badgers, detailed necropsy, histopathological examination of suspect tissue and bacteriological examination were carried out on tissues from all carcasses. Tuberculosis was diagnosed in 42 (51%) of these badgers. Gross lesions (confirmed by histopathological examination) were found in 28 (67%) of the total infected, and the remaining 14 (33%) were identified by isolation of *Mycobacterium bovis* from tissues of badgers which had no gross lesions of the disease. Forty per cent of the lesions identified were in the thoracic cavity, 29% in the carcass (skin, popliteal and prescapular lymph nodes), 16% in the tissues and/or lymph nodes of the head and neck, and 15% in the abdominal cavity.

INTRODUCTION

Necropsy with bacteriological examination of tissues is the most effective method of diagnosing tuberculosis in badgers (Pritchard *et al.* 1986). Effective skin and blood tests for the diagnosis of TB in live badgers are still under trial (Nolan and Goodger 1993). The testing of urinary, faecal or salivary excretions has only a limited usefulness since the animals must be shedding the organism at the time of sampling to be identified.

In Ireland, the carcasses of badgers trapped under licence are submitted to the veterinary laboratories of the Department of Agriculture and Food for examination for evidence of tuberculosis. Necropsy is performed on each carcass and tissue suspected of being diseased is taken for histopathological examination. Bacteriological examination is not routinely carried out. Investigations in Britain undertaken at the Gloucestershire Veterinary Investigation Centre between 1972 and 1979 and reported by Zuckerman (1980) have found that the diagnostic rate of tuberculosis in badgers was substantially increased when bacteriological examination was carried out on tissues from badgers with no gross lesions.

The aims of this study were (1) to determine the incidence of TB in a group of badgers, (2) to determine the diagnostic rate for necropsy, (3) to determine the diagnostic rate when bacteriological examination of tissues was also undertaken, and (4) to document the locations of TB lesions found in badgers.

MATERIALS AND METHODS

All badgers submitted to the Veterinary Research Laboratory at Abbotstown from mid-October 1988 until the end of January 1989 and to the Regional Veterinary Laboratory at Athlone in June 1989 were included in the study. These badgers had been trapped under licence in areas with a high prevalence of bovine tuberculosis.

A detailed examination of each carcass was carried out by the author and particular attention was paid to tissues listed in Table 1. Each tissue was examined visually for evidence of gross lesions of tuberculosis and the lungs were palpated. The tissues were then serially sliced and the cut surfaces examined. The locations of lesions suspected of being tuberculous were recorded, and samples of these tissues were taken and subjected to histopathological and bacteriological examinations. If no suspect lesion was noted, a pool of tissue was formed by taking samples of lymph nodes from each region of the body (Table 1) and this tissue pool was subjected to bacteriological examination.

Table 1. Anatomic location of lymphoid tissues and organs examined during necropsy. L.N. = Lymph node.

BODY REGION	TISSUE EXAMINED	BODY REGION	TISSUE EXAMINED
Head and neck	Pharyngeal L.N. Parotid L.N. Submandibular L.N.	**Thoracic cavity**	Lung tissue Mediastinal L.N. Bronchial L.N. Liver
Abdominal cavity	Kidney Spleen Hepatic L.N. Mesenteric L.N.	**Carcass**	Skin Prescapular L.N. Popliteal L.N.

Tissue samples for histopathological examination were prepared by fixing in 10% buffered formalin, embedding in paraffin wax, slicing at 5-7 micrometre thickness and staining with both haematoxylin and eosin and Ziehl Neilsen stains. The tissue samples for bacteriological examination were suspended in 5% oxalic acid, macerated with a stomacher, washed in normal saline before the decontaminated deposits were innoculated onto slopes of Stonebrink's medium, and incubated at 37°C for six weeks. TB was diagnosed by (1) isolating and typing *M. bovis* or (2) identifying the histopathological lesion, including observation of acid-fast bacilli, as described by Gallagher *et al.* (1976).

RESULTS

Eighty-three badgers were examined in this study and 42 (51%) showed evidence of tuberculosis (Table 2). Gross lesions were evident in 28 (67%) of these and *M. bovis* was isolated on culture from the remaining 14 (33%) which had no visible lesions (Table 2). Thus there was a significant increase in the incidence of TB when tissues were examined by both necropsy and culture as opposed to necropsy only ($X^2 = 4.84$, $P < 0.05$). There was, however, no significant difference between the incidence of gross lesions (28/83) and the proportion (37/83) of culture-positive badgers ($X^2 = 2.05$, $P > 0.05$). This is because *M. bovis* could not be cultured from all animals with lesions and was identified from some with no visible lesions. *M. bovis* was isolated from 23 (82%) of the 28 badgers with lesions. Overall, 23/83 (27.7%) specimens proved positive by both assay methods. This is significantly less than the overall incidence of 42/83 (50.6%), as assessed by either culture or necropsy ($X^2 = 17.4$, $P < 0.001$).

Table 2. Necropsy results for 83 badgers examined following capture under licence in areas with a high prevalence of bovine tuberculosis.

(a) Total no. of animals examined No. infected with TB	83 42 (50.6%)
(b) Total no. infected with TB No. with gross lesions of TB No. with no visible lesions from which *M. bovis* was isolated	42 28 (66.7%) 14 (33.3%)
(c) No. with gross lesions of TB No. of these from which *M. bovis* was isolated	28 23 (82.1%)

Table 3. The distribution of gross lesions of tuberculosis in 17 badgers affected at one site only.

INFECTED AREA	NUMBER OF ANIMALS	INFECTED AREA	NUMBER OF ANIMALS
Pharyngeal lymph node	1	Lung tissue	1
Submandibular lymph node	1	Bronchial lymph node	2
Parotid lymph node	1	Mediastinal lymph node	1
Liver	1	Prescapular lymph node	3
Spleen	1	Popliteal lymph node	4
Mesenteric lymph node	1		

The locations of gross TB lesions are documented in Tables 3 and 4. Lesions were found at 15 sites in the body. Forty per cent of the total lesions found occurred in the thoracic cavity, 29% in the carcass, 16% in the head and neck region, and 15% in the abdominal cavity (Table 5). The distribution of lesions was non-random ($X^2 = 12.48$, $P < 0.005$) and the thoracic cavity and carcass had significantly more lesions than the head and neck region and abdominal cavity. There were 17 badgers with a TB lesion at one site only and 11 with lesions at two or more sites.

Table 4. The distribution of tuberculous lesions (*) in 11 badgers with infection observed at more than one site.

INFECTED AREA	BADGER NUMBER										
	18	19	20	21	22	23	24	25	26	27	28
Pharnygeal L.N.	-	-	*	-	-	-	-	-	-	*	**
Parotid L.N.	-	-	-	*	-	-	-	-	-	-	-
Submandibular L.N.	-	-	-	-	-	-	-	-	-	*	-
Lung tissue	*	*	**	-	-	-	-	*	-	*	-
Bronchial L.N.	*	-	-	*	-	*	*	-	-	**	**
Mediastinal L.N.	-	-	-	-	-	-	*	-	-	-	*
Pericardium	-	-	-	-	-	-	-	-	-	-	*
Pleura	-	-	-	-	-	-	-	-	-	-	*
Liver	-	-	-	-	-	-	-	-	-	*	-
Kidney	-	-	-	-	-	-	-	*	-	*	*
Spleen	-	-	-	-	-	-	-	-	-	*	-
Prescapular L.N.	-	-	-	-	-	-	-	*	*	-	-
Popliteal L.N.	-	-	-	-	**	*	*	-	-	**	-
Skin	-	-	-	-	-	-	-	-	*	-	-

Table 5. Distribution of gross lesions throughout the regions of the body.

BODY REGION	LESION AT 1 SITE ONLY	LESION AT > 1 SITE	TOTAL (%)
Head and neck	3	6	9 (16.4%)
Thoracic cavity	4	18	22 (40.0%)
Abdominal cavity	3	5	8 (14.5%)
Carcass	7	9	16 (29.1%)

On gross examination, lesions of tuberculosis manifested themselves as follows. In the lungs, pale grey spherical nodules were encountered which, in advanced cases, coalesced to form large tuberculous masses. Miliary (seed-like) lesions were found on the pleurae and pericardium in one badger (Table 4). In the liver and kidney, TB lesions occurred as discrete pinhead areas and were also pale grey in colour. The cut surface of the TB-infected skin lesion was granular in appearance and was accompanied by purulent exudate. Affected lymph nodes were frequently enlarged and tuberculous lesions appeared as areas of caseation (yellowy cheese-like material). Calcification was seen infrequently in lymph nodes and it manifested as white chalky material. Parasitism was the principal differential diagnosis. On histopathological examination, the typical TB lesion had a caseous centre which contained epithelioid cells. This was surrounded by macrophages with lymphocytes on the periphery. Fibroblastic reaction was slight and giant cells were not seen. Calcification was noted infrequently. Some lesions contained large numbers of acid-fast bacilli while others contained very few.

DISCUSSION

The incidence of tuberculosis (51%) in the badgers in this study was high. There were no gross lesions of TB in 33% of the infected animals. In Gloucestershire, Gallagher *et al.* (1976) reported that 14% of infected badgers had no gross evidence of the disease. Zuckerman (1980) reported that the proportion of infected badgers with no gross lesions of TB examined at the Gloucestershire Veterinary Investigation Centre increased from 20% for the years 1972-5 to 40%, 40% and 44% for the years 1977, 1978 and 1979 respectively. In 1976, 67% of infected badgers had no gross lesions of TB. In this study, *M. bovis* was isolated from 82% of badgers with gross lesions of TB, which was a higher recovery rate of the organism than that reported by Pritchard *et al.* (1986), who recovered *M. bovis* by cultural methods from 5 (56%) of 9 badgers with visible lesions of TB.

Tuberculous lesions occurred in the lungs or pulmonary lymph nodes in 13 (46%) of the 28 badgers with gross evidence of the disease. This suggests that inhalation was a very important route of infection for these badgers. Gallagher *et al.* (1976) found lesions in the lungs or pulmonary lymph nodes in 22 (71%) out of 31 badgers with gross lesions of tuberculosis. Cheeseman *et al.* (1989) reported that infection was acquired via the respiratory route in 23 (82%) of 28 badgers examined.

Lesions of TB were found in either the prescapular or popliteal lymph nodes or skin in 13 badgers, 8 of which did not have a lesion in any other region of the body (Tables 3 and 4). Since these lymph nodes drain superficial areas of the body, it is suggested that biting by another badger was the route of infection for these 8 badgers, which represent 29% of the total (28) with gross lesions. Gallagher and Nelson (1979) reported that TB infection had been introduced by a wound inflicted by another badger in 15 (56%) of 27 badgers which had died of the disease.

Renal involvement was found in only 3 badgers (11%), each of which had TB lesions at two or more other sites. In comparison, Gallagher *et al.* (1976) found renal involvement in 6 (19%) of 31 badgers with gross lesions of TB, whilst Little *et al.* (1982) found tuberculous lesions in the kidneys of 9 (75%) out of 12 infected badgers examined.

The widespread distribution of lesions (Tables 3 and 4) indicates that TB can affect most lymphoid tissue and organs in the badger. Since 40% of the total lesions occurred in the thoracic cavity (Table 5), this is the region of the body in which the pathologist is most likely to find evidence of TB. The carcass lymph nodes (prescapular and popliteal) and the skin accounted for 29% of the lesions, with an additional 16% and 15% of gross lesions recorded in the head and neck region and abdominal cavity respectively. These findings indicate that there was a high incidence of TB in the badger populations from which these animals came, that post-mortem bacteriological examination substantially increased the diagnostic rate of infected badgers, that inhalation and biting are common routes of TB infection for badgers, and that TB lesions may be found in most lymphoid routes of TB infection for badgers, and that TB lesions may be found in most lymphoid tissue and organs in the badger.

ACKNOWLEDGEMENTS

The author wishes to thank the staff of the tuberculosis and histology sections of the Central Veterinary Laboratory, Abbotstown, for carrying out bacteriological examinations and for preparing slides for histopathological examination, and Ms Sharon McAuley for typing the manuscript.

REFERENCES

CHEESEMAN, C.L., WILESMITH, J.W. and STUART, F.A. 1989 Tuberculosis: the disease and its epidemiology in the badger: a review. *Epidemiology and Infection* **103**, 113-25.

GALLAGHER, J. and NELSON, J. 1979 Cause of ill health and natural death in badgers in Gloucestershire. *Veterinary Record* **105**, 546-51.

GALLAGHER, J., MUIRHEAD, R.H. and BURN, K.J. 1976 Tuberculosis in wild badgers (*Meles meles*) in Gloucestershire: pathology. *Veterinary Record* **98**, 9-14.

LITTLE, T.W.A., NAYLOR, P.F. and WILESMITH, J.W. 1982 Laboratory study of *Mycobacterium bovis* infection in badgers and calves. *Veterinary Record* **111**, 550-7.

NOLAN, A. and GOODGER, J. 1993 Diagnosis of tuberculosis in live badgers. In T.J. Hayden (ed.), *The badger,* 128-30. Dublin. Royal Irish Academy.

PRITCHARD, D.G., STUART, F.A., WILESMITH, J.W., CHEESEMAN, C.L., BREWER, J.I., BODE, R. and SAYERS, P.E. 1986 Tuberculosis in East Sussex. III. Comparison of post-mortem and clinical methods for the diagnosis of tuberculosis in badgers. *Journal of Hygiene, Cambridge* **97**, 27-36.

ZUCKERMAN, LORD 1980 *Badgers, cattle and tuberculosis.* London. HMSO.

In: Hayden, T.J. (ed.) 1993 *The badger*, 123-127. Dublin. Royal Irish Academy.

STRUCTURAL ORGANISATION OF THE BADGER KIDNEY

A.S. Elgadi and T.J. Hayden†

Mammal Research Group, Zoology Department, University College, Dublin

†To whom correspondence should be addressed.

INTRODUCTION

If badgers transmit tuberculosis to cattle, the most likely routes are by direct contact between cattle and infectious badgers or their excretory products. The tuberculosis lesions are frequently detected in the kidneys of badgers (Fagan 1993), and in all but a few individuals the number of bacilli in urine may be as high as 300,000 mL^{-1} (MAFF 1979). Accordingly the pattern of urine deposition by the badgers on pasture (Brown *et al.* 1993) and the reaction of cattle to contaminated pasture are under active examination (Benham and Broom 1991). Although the histopathology of the TB lesions in the kidneys of badgers is routinely examined, rather less attention has been paid to the effect of lesions on the structural organisation of the kidney. Indeed, there has been little or no investigation of the ultrastructure of the normal badger kidney. We here present a preliminary report on the structural organisation of this organ.

METHODS

Healthy kidneys from adult badgers were provided by the Regional Veterinary Laboratory, Athlone. External dimensions were measured with a vernier calliper. Kidney size, relative medullary thickness (RMT) and percent medullary thickness (PMT) were determined (Sperber 1944; Heisinger and Breitenbach 1969). Casts of the renal pelvis were made by infusing latex via the ureter and then macerating the kidneys in HCl. The general vascular pattern was determined by infusion of diluted latex and vinyl acetate via the renal artery/vein. The kidneys were then macerated (Pfeiffer *et al.* 1960). The pattern of the fine blood vessels was revealed by injecting India ink before dehydration, clearing and sectioning the tissue at 50µm using a CO_2 freezing microtome. Kidneys for histological studies were fixed in Bouin's solution, washed, dehydrated, cleared in cedarwood oil and embedded in paraplast. Serial sections (8µm) were stained with Haematoxylin and Eosin (Humason 1979).

RESULTS

The kidney of the Eurasian badger is a flattened, more or less bean-shaped organ covered by a fibrous capsule. It is a crested kidney in which several collecting ducts open

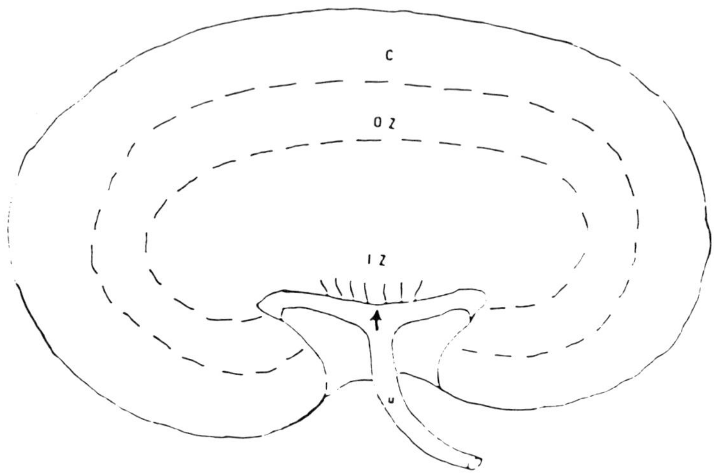

Fig. 1. Diagram showing the crested kidney of the badger (arrow) and zonation. C=cortex, O.Z.=outer medullary zone, I.Z.=inner medullary zone, U=ureter.

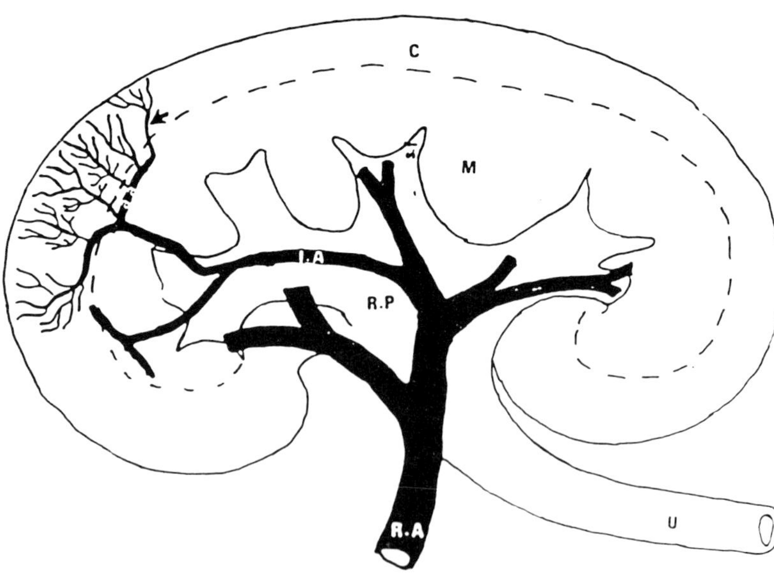

Fig. 2. Diagram showing the arrangement of the arterial blood supply and the renal pelvis in the badger kidney. R.A.=renal artery, I.A.=interlobar artery, a.a.=arcuate artery (arrow showing cortical radial artery), S.F.=specialised fornix, R.P.=renal pelvis, U=ureter, C=cortex, M=Medulla.

on a ridge and shows medullary organisation (Fig. 1). In adults the mean length, width and depth are 56.24mm, 27.53mm and 30.36mm respectively. The mean weight of the paired kidneys is about 60g (0.5-0.75% of the total body weight of the animal). The maximum urine osmolarity of mammals is linearly related to the degree of medullary development expressed either as RMT (R = 0.86) or PMT (R = 0.88) The RMT for badger kidney is 3.41, the PMT is 68.33. These correspond to a theoretical maximum osmolarity of the urine of 1861 and 2691/mOsm/L respectively.

The renal artery upon entering the hilus divides into nine interlobar arteries. These interlobar arteries radiate from the hilus along the external surface of the pelvic wall. At the cortical medullary boundary, the interlobar form the arcuate arteries (Fig. 2). Cortical radial arteries branch from the convex side of the arcuate and penetrate the cortex. The veins follow the same pattern as the arteries.

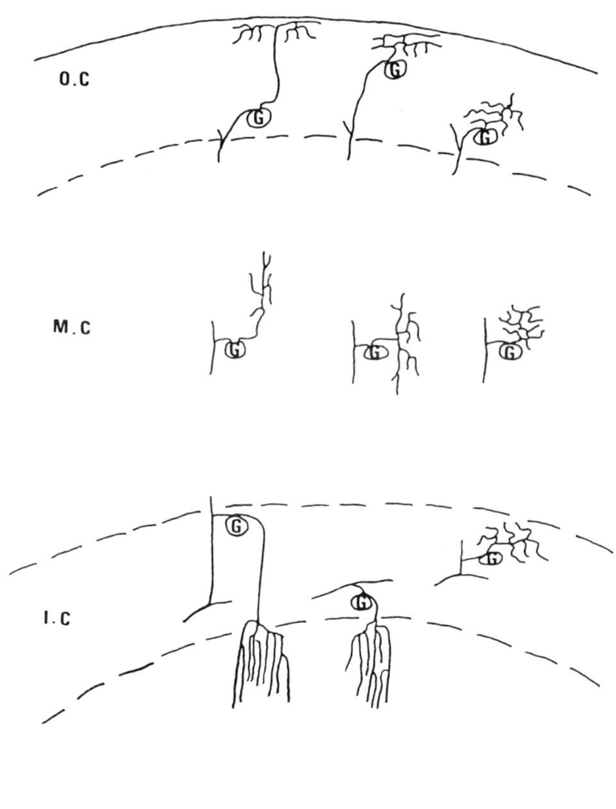

Fig. 3. The three anatomical zones of the cortex. O.C.=outer cortex, M.C.=mid-cortex, I.C.=inner cortex, and G=glomerulus.

The nephrons in the badger are all characterised by loops of the long type, which descend into the inner medulla before they turn. The glomerular density appears to be high. Three different types of efferent arterioles are distinguished, namely sub-scapular, mid-cortical and inner-cortical (Fig. 3). The sub-scapular type breaks up on the kidney surface forming the stellate vessels. The medulla is supplied by efferent arterioles from the inner cortex.

The vascular bundles found in the badger are of the simple type composed mainly of descending and ascending vasa recta. This can be considered as a counter-current exchange system between arteriole and venous vasa recta.

The badger possesses a renal pelvis composed of specialised fornices (type II in the terminology of Pfeiffer 1968) that project deep into the medullary tissue (Fig. 2). The pelvic wall is well supplied with smooth muscle fibres, contractions of which in a pelvis of this type cause urine to reflux into the deep fornices in the upper pelvis (Atherton *et al.* 1972; Boyarsky and Labay 1972). Thus the outer and inner medullary area are bathed in urine, so that most of the parenchyma is in close proximity to the post-tubular urine. This provides more direct access to urine for bacilli released from kidney lesions in comparison with the normal urinary route through the nephrons and collecting ducts. We suggest that this is a contributory factor to the dissemination of bacilli in large numbers. It does not, however, allow an assessment of the risk this contamination poses to badgers or other animals.

REFERENCES

ATHERTON, J.C., GREEN, R., THOMAS, S. and WOOD, J.A. 1972 Time course of changes in renal tissue and urinary composition after cessation of constant infusion of lysine vasopressin in the conscious hydrated rat. *Journal of Physiology* **222**, 583-95.

BENHAM, P.F.J. and BROOM, D.H. 1991 Responses of dairy cows to badger urine and faeces in pasture with reference to bovine tuberculosis transmission. *British Veterinary Journal* **145**, 226-41.

BOYARSKY, S. and LABAY, P. 1972 *Uretheral dynamics: pathophysiology, drugs and surgical implications.* Baltimore. Williams and Wilkins.

BROWN, J.A., HARRIS, S. and CHEESEMAN, C.L. 1993 The development of field techniques for studying potential modes of transmission of bovine tuberculosis from badgers to cattle. In T.J. Hayden (ed.), *The badger,* 139-53. Dublin. Royal Irish Academy.

FAGAN, J. 1993 Tuberculosis in badgers in Ireland: pathology. In T.J. Hayden (ed.), *The badger,* 117-22. Dublin. Royal Irish Academy.

HEISINGER, J.F. and BREITENBACH, P.R. 1969 Renal structural characteristics as indices of renal adaptation for water conservation in the genus *Sylvilagus. Physiological Zoology* **42**, 160-72.

HUMASON, G.L. 1979 *Animal tissue techniques* (4th edn). San Francisco. W. H. Freeman and Co.

MAFF 1979 *Bovine tuberculosis in badgers third report.* Ministry of Agriculture, Fisheries and Food. London. HMSO.

PFEIFFER, E.W. 1968 Comparative anatomical observations of the mammalian renal pelvis and medulla. *Journal of Anatomy* **102**, 321-31.

PFEIFFER, E.W., NUNGESSER, W.C., IVERSON, D.A. and WALLERIUS, J.F. 1960 The renal anatomy of the primitive rodent *Aplodontia rufa* and a consideration of its functional significance. *Anatomical Record* **137**, 227-35.

SPERBER, I. 1944 Studies on the mammalian kidney. *Zoologiska Bidrag Fran Uppsala* **22**, 249-431.

In: Hayden, T.J. (ed.) 1993 *The badger*, 128-130. Dublin. Royal Irish Academy.

DIAGNOSIS OF TUBERCULOSIS IN LIVE BADGERS

A. Nolan† and J. Goodger

Central Veterinary Laboratory, Weybridge, Surrey KT15 3NB, UK

† To whom correspondence should be addressed.

ABSTRACT

An antigen which has *Mycobacterium bovis*-specific epitopes has been identified as immunodominant in the antibody responses of badgers to infection with *M. bovis*. A kit using a purified form of this antigen in an indirect ELISA system, suitable for field use, is described.

INTRODUCTION

Clinical diagnosis of *M. bovis* infection in badgers by sampling of faeces, urine and tracheal aspirates, followed by standard culture procedures, has been shown to be a poor indicator of disease (Pritchard *et al.* 1986). In addition, it is an impractical method of establishing disease status for control purposes because of the time necessary for incubation of cultures. An immunodiagnostic method of detecting infection would appear to be an alternative of choice; however, skin tests have been shown to be of no practical value (Pritchard *et al.* 1986), and the alternative of measuring antibody to *M. bovis* has been beset with problems owing to the lack of specificity. Relatively high levels of cross-reactive antibodies compared with specific antibodies were shown to be present in two studies of infected badgers (Morris *et al.* 1979; Mahmood *et al.* 1987). Identification of antigens with specificity for *M. bovis* is therefore a prerequisite for improving the prospect of immunodiagnosis. This paper describes the use of immunoblotting of mycobacterial antigens to identify such an antigen, and summarises the research which has led to the development of an indirect ELISA using purified antigen for a 'field test' for *M. bovis* infection in badgers.

MATERIALS AND METHODS

Immunoblotting of badger sera

Sonicated extracts of *M. bovis* were separated under reducing conditions (2-mercaptoethanol) in 12.5% SDS-PAGE gels using a discontinuous Tris-HCl buffer system (Laemmli 1970). Antigens were transferred to nitrocellulose sheets (Biorad, Transblot membrane) by electroblotting using a tris-glycine-methanol buffer (Towbin *et al.* 1979). Assays were carried out at room temperature on an orbital shaker. Non-specific sites were blocked with 3% (w/v) bovine serum albumin (BSA) for one hour. Nitrocellulose sheets were incubated for one hour with antisera diluted 1/2, 1/5 or 1/20 in 1% (w/v) BSA in phosphate-buffered saline (0.01M, pH 7.2) and 0.05% Tween 20 (PBST + BSA). After washing, blots were incubated with monoclonal antibody to badger immunoglobulins (Igs) for one hour. After washing, blots were incubated in affinity-purified anti-mouse Ig biotinylated species-specific $F(Ab')_2$ fragment (Amersham) in 1% (w/v) BSA-PBST for 90 minutes, washed in PBST and further incubated in Streptavidin biotinylated horseradish peroxidase complex (Amersham) for 30 minutes. After washing in PBS, labelled sites were visualised using 0.05% chloronaphthol with 0.015% hydrogen peroxide. Reactions were stopped by washing in water.

RESULTS AND DISCUSSION

Serum from two experimentally infected badgers identified a major band at 26kDa and two minor bands at 22kDa and 18kDa respectively. Serum from six naturally infected badgers and two badgers from areas free of tuberculosis and which had no clinical or bacteriological signs of infection were assayed at 1/5 and 1/20 dilutions of serum. Four of the six infected badger sera showed binding to the 26kDa band and to a faint blurred region between 40 and 60kDa. Three of the four also showed binding to the 26kDa band at a 1/20 dilution of serum. None of the assays with negative sera showed any activity. Subsequently, sera from 26 infected and 17 uninfected badgers were assayed at a 1/2 dilution of serum in an attempt to improve sensitivity. Sixteen of the 26 infected badger sera showed activity of various levels in the 26kDa region. Serum from 5 of the 17 uninfected badger sera also showed activity in this region. The binding was faint in several cases and the banding was not as clear as in the previous assays.

Studies using polyclonal rabbit antisera and monoclonal antibodies have identified the 26kDa antigen as an immunodominant antigen with *M. bovis*-specific epitopes (Nolan, unpubl. data). The data from the blotting of the badger sera indicate that this antigen is also dominant in generating antibodies in *M. bovis*-infected badgers.

Using the monoclonal antibody which binds to the 26kDa antigen and which has specificity for *M. bovis*, a blocking ELISA system has been developed at the Central Veterinary Laboratory. The assay uses a crude antigen and relies upon the binding of the MAb to an *M. bovis* epitope to generate the specificity of the assay. The stringency of the system has been set to give a very high specificity (> 98%) and the sensitivity has been demonstrated to be linked to the pathogenesis of the disease. Unfortunately, because of the nature of the target antigen, the system has proved insufficiently robust to be considered for field use, owing to a variety of physiochemical conditions (Nolan, unpubl. data).

Purification of the antigen has allowed the development of an indirect ELISA system which has now been adapted for field use (Nolan, unpubl. data). The system requires certain steps to be carried out at 37°C and a mobile laboratory has been constructed to test the system in the field. The procedure for using the test is as follows: badgers are trapped in the conventional way, transferred to carrying cages and transported to the mobile laboratory site. Following anaesthetisation using ketamine hydrochloride (Vetalar, Parke-Davis) (Mackintosh *et al.* 1976), a whole blood sample is collected, from the jugular vein, into a heparinised vacutainer. In order to speed the process, the plasma is separated from red cells by a brief centrifugation step in a microcentrifuge. The plasma is diluted 1/10 in the appropriate diluent and dispensed into wells of the precoated ELISA plate. The system is controlled using positive, negative, cutoff and blank controls which are supplied freeze-dried and are reconstituted and treated in the same way as the plasma. The plate is incubated for 20 minutes at 37°C in a shaking incubator. After washing, a peroxidase-conjugated monoclonal antiglobulin (supplied freeze-dried and reconstituted before use) is added to the wells and incubated for a further 20 minutes. The plate is washed and substrate (tetramethylbenzidine, Cambridge Veterinary Sciences) is added and the plate incubated for ten minutes. The test can be read visually (blue) or stopped with citric acid and read visually (yellow) or at 450nm.

A trial has just begun in the south-west of England to investigate the practicalities of using the test and to monitor factors which may affect its performance. Animals which are to be tested will be selected from routine control operations, and post-mortem examination and culture will provide additional data for the evaluation of the test performance under field conditions. The laboratory evaluation has so far proved favourable, with a very high specificity (> 98%), and correlation with the monoclonal blocking assay is also very high. Data showing the dependence of the development of specific antibody responses of badgers on the degree of infection with *M. bovis* will be published in more detail elsewhere (Nolan, unpubl. data). The data suggest that the antibody response is closely associated with the pathogenesis of the disease, and evidence has suggested a difference in the pathogenesis of disease in young animals which renders the antibody assays more sensitive in cubs overall (Nolan, unpubl. data).

REFERENCES

LAEMMLI, U.K. 1970 Cleavage of the structural proteins during the assembly of the head of the bacteriophage T4. *Nature* **227**, 680-5.

MACKINTOSH, C.G., MACARTHUR, J.A., LITTLE, T.W.A. and STUART, P. 1976 The immobilisation of the badgers (*Meles meles*). *British Veterinary Journal* **132**, 609-14.

MAHMOOD, K.H., ROOK, G.A.W., STANFORD, J.L., STUART, F.A. and PRITCHARD, D.G. 1987 The immunological consequences of challenge with bovine tubercule bacilli in badgers (*Meles meles*). *Epidemiology and Infection* **98**, 155-63.

MORRIS, J.A., STEVENS, A.E., STUART, P. and LITTLE, T.W.A. 1979 A pilot study to assess the usefulness of ELISA in detecting tuberculosis in badgers. *Veterinary Record* **104**, 14.

PRITCHARD, D.G., STUART, F.A., WILESMITH, J.W., CHEESEMAN, C.L., BREWER, J.I., BODE, R. and SAYERS, P.E. 1986 Tuberculosis in East Sussex. III. Comparison of the post mortem and clinical methods for the diagnosis of tuberculosis in badgers. *Journal of Hygiene (Cambridge)* **97**, 27-36.

TOWBIN, H., STACHELIN, T. and GORDON, J. 1979 Electrophoretic transfer of proteins from polyacrylamide gels to nitrocellulose sheets: procedure and some applications. *Proceedings of the National Academy of Science, USA* **76**, 4370-454.

In: Hayden, T.J. (ed.) 1993 *The badger*, 131-138. Dublin. Royal Irish Academy.

BADGERS AND BOVINE TUBERCULOSIS: EPIDEMIOLOGY AND CASE STUDIES

P. Flanagan

District Veterinary Office, Tullamore, Co. Offaly

ABSTRACT

There is an increasing awareness and acceptance that the badger is very influential in bovine TB in Ireland by seeding infection into the national herd. However, the importance of badger-to-cattle transmission, relative to cattle-to-cattle transmission, has not been quantified. The two case studies outlined in this paper highlight the role of the badger in TB breakdowns in herds.

INTRODUCTION

Tuberculosis (TB) is an infectious disease in bovines and other animals caused by *Mycobacterium bovis*. In Ireland, a compulsory TB eradication scheme based on a 'test and slaughter' policy commenced in the 1950s. All bovines are subjected to the tuberculin test (single intradermal comparative test). Animals which react positively (reactors) are removed for immediate slaughter. On completion of the test, the veterinary surgeon reports the results of the test to the local district veterinary office (DVO).

DVO procedure

On receipt of a report of a test on which reactors were disclosed, the following procedures are undertaken at the DVO.

(1) The report is examined by a veterinary inspector (VI) to ensure that the appropriate interpretation was applied by the testing veterinarian.
(2) The outbreak is classified, depending on its severity, into a category A, B or C type breakdown. Category A breakdown is the most severe and applies to a herd where 10% or more of the animals fail the tuberculin test.
(3) A notice restricting the movement of animals into, or out of, the herd (except under permit for slaughter), together with information on the removal of reactor(s) for slaughter, the disinfection of the premises (where relevant) and the relevant period of restriction, is issued to the herd-owner.
(4) An epidemiological investigation to establish the most likely source of infection is initiated (Griffin *et al.* 1992).

Epidemiology

Epidemiology is defined as the study of the factors which cause or influence a disease. Table 1 outlines the possible sources of infection (TB). The investigation begins in the DVO where the VI examines the herd file, noting details of previous restrictions of the herd and of any purchased animal that was deemed a reactor. Where a bought-in animal fails a test, reference is made to the herd of origin to establish the likelihood of 'bought-in' infection. The holding is visited by a VI and all the lands pertaining to this herd are mapped and details recorded regarding:

type of farming enterprise (dairy/suckler/store/beef),
stock replacement (self-contained/buy in),
fragmentation of holding,
grazing management (intensive/extensive),
water sources (river/stream/pond/well),
slurry management (contractor/own equipment),
TB status of herd-owner and staff,
the evidence of wildlife (badgers and/or deer) on the farm or in the area.

The foregoing information is carefully analysed with a view to establishing the most likely source of the outbreak. In this paper two case studies are described where the epidemiological investigations strongly indicated an involvement of badgers in the outbreaks of TB in cattle.

Table 1. Possible sources of tuberculosis.

A.	Lateral (bovine to bovine) spread from adjoining herds.
B.	Bought-in animals.
C.	Residual infection from a previous outbreak.
D.	Mechanical vector, i.e. slurry-spreader or trailer.
E.	Wildlife (badgers/deer).
F.	Human

CASE STUDIES

Case 1

Case 1 concerns a farm (Farm 1) in the north-west of County Offaly. The herd was maintained in two separate groups which were grazed extensively on a fragmented holding. All animals were outwintered. All drinking water was piped to troughs placed in the grazing area. Seven bullocks (group 1) were grazed on fields marked a, b, and c (Fig. 1), while the remaining 35 animals (group 2) were grazed on that portion of the farm marked d (Fig. 1). Neither group grazed on e, f or g during this time. Tuberculin test summaries for herd 1 and its nine contiguous herds (2-10) are outlined in Table 2. All these herds passed their respective tests during the period December 1985 to February 1986. On 6 May 1986 a dead badger was found on the land where group 1 grazed (marked on Fig. 1). Evidence of trampled grass, bovine hoof marks and fresh bovine faeces in the area immediately surrounding the badger carcass suggested the probability of contact between group 1 and the dying/dead badger.

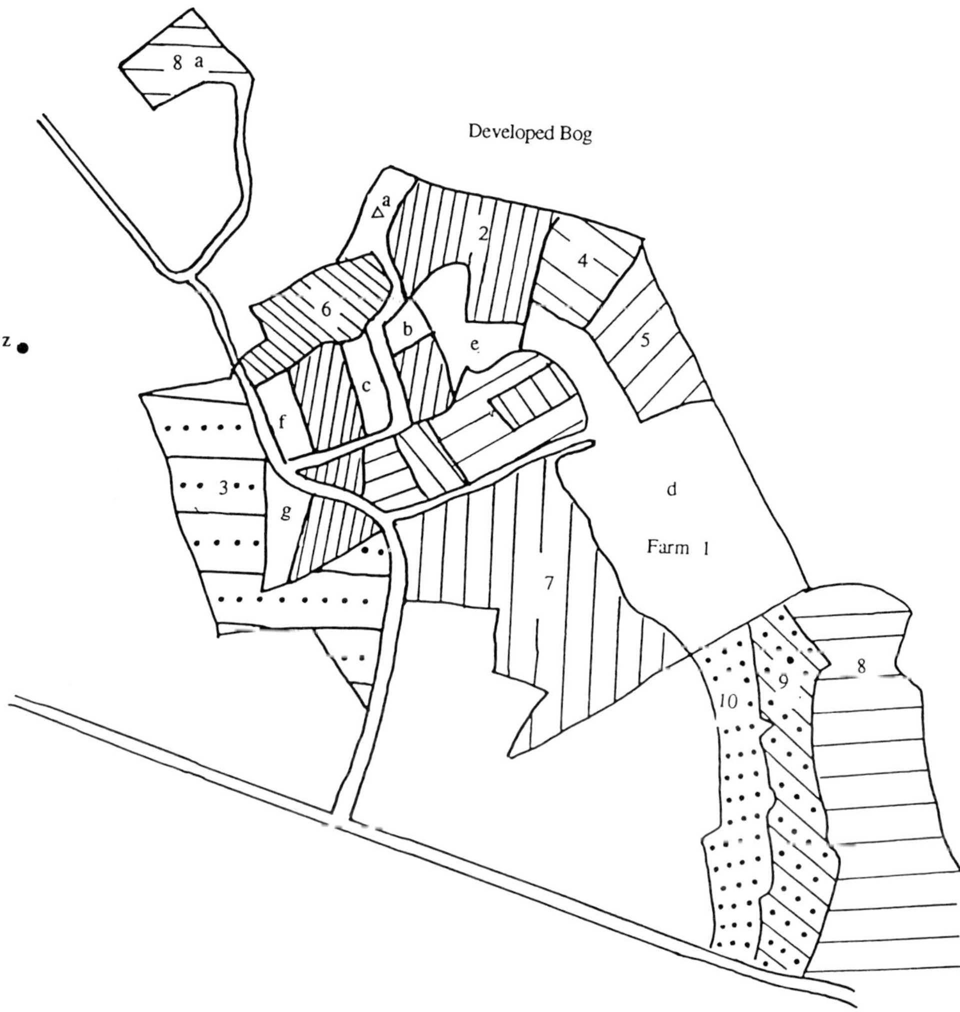

Fig. 1. Geographical relationship between Farm 1 and adjacent herds (numerals). Pasture areas in Farm 1 are identified by letter codes. A dead badger (Δ) was found in field a and the nearest sett is indicated at Z.

Following confirmation of generalised TB in the badger carcass at the regional veterinary laboratory, group 1 was tested on 13 June 1986 and four reactors were disclosed, one showing visible lesions at slaughter. Group 2 was tested two weeks later, and a further reactor was disclosed which also showed visible lesions at slaughter. Contiguous herds 4 and 8 produced reactors on their subsequent tests: two visible lesion (VL) reactors were identified in herd 4, while two NVL (no visible lesion) reactors were disclosed in herd 8. The reactors in herd 4 were outwintered beside group 2. The reactors in herd 8 were housed from November 1985 to mid-May 1986, after which time they were turned out to the outfarm marked 8a (top left-hand corner on Fig. 1). The remaining three animals in group 1 which passed the test on 13 June also passed two further tests, and when they were eventually slaughtered no visible TB lesions were seen on post-mortem examination.

Table 2. Summary of tuberculin tests for herd 1 and nine contiguous herds (2-10) between November 1985 and October 1986. (C = clear herd test; numeral = number of reactors on test, VL = visible lesions, NV= non-visible lesions, * = time when TB-positive badger found on farm.)

Herd	Nov.	Dec.	Jan.	Feb.	Mar.	Apr.	May	Jun.	Jul.	Aug.	Sep.	Oct.
1	-	-	C	-	-	-	*	5	-	C	-	-
2	-	-	C	-	-	-	-	-	C	-	-	-
3	-	-	C	-	-	-	-	-	C	-	-	-
4	-	-	-	C	-	-	-	-	2 VL	-	C	-
5	-	-	-	C	-	-	-	-	C	-	-	-
6	-	-	-	C	-	-	-	-	-	C	-	-
7	-	-	-	C	-	-	-	-	-	C	-	-
8	-	-	C	-	-	-	-	-	2 NV	-	C	-
9	-	C	-	-	-	-	-	-	-	C	-	-
10	-	-	-	C	-	-	-	-	-	-	-	-

Badgers

Six badgers were snared in the area under licence in August 1986 and one was TB positive on post-mortem examination at the regional veterinary laboratory. The sett where this positive badger was captured is marked Z (Fig. 1).

Conclusions

1. As group 1 and the animals from contiguous herd 8 were totally isolated from either group 2 or contiguous herd 4, it is extremely unlikely that lateral spread between herds was of any importance in this breakdown.
2. Herds 1, 4 and 8 had very good testing histories prior to this outbreak, and none disclosed reactors subsequently. All the reactors passed the test in the period December 1985 to February 1986. Therefore there are no grounds for suspecting that the likely source of infection might be residual infection or bought-in animals.
3. As all watercourses in Farm 1 were fenced off, and all cattle were outwintered, there was nothing to suggest that the infection was mechanically introduced.
4. There was no evidence of human involvement in the breakdown.

Based on the foregoing, it would appear that the dead badger was the most likely source of the outbreak.

Case 2

Case 2 concerns a TB breakdown in an intensive dairy herd. This unfragmented farm (Farm A), which is located in east Offaly, is well suited to grazing. The southern end of the farm consists of reclaimed bog. When this reclamation was being carried out, the herd-owner decided not to remove a large badger sett (X on Fig. 2).

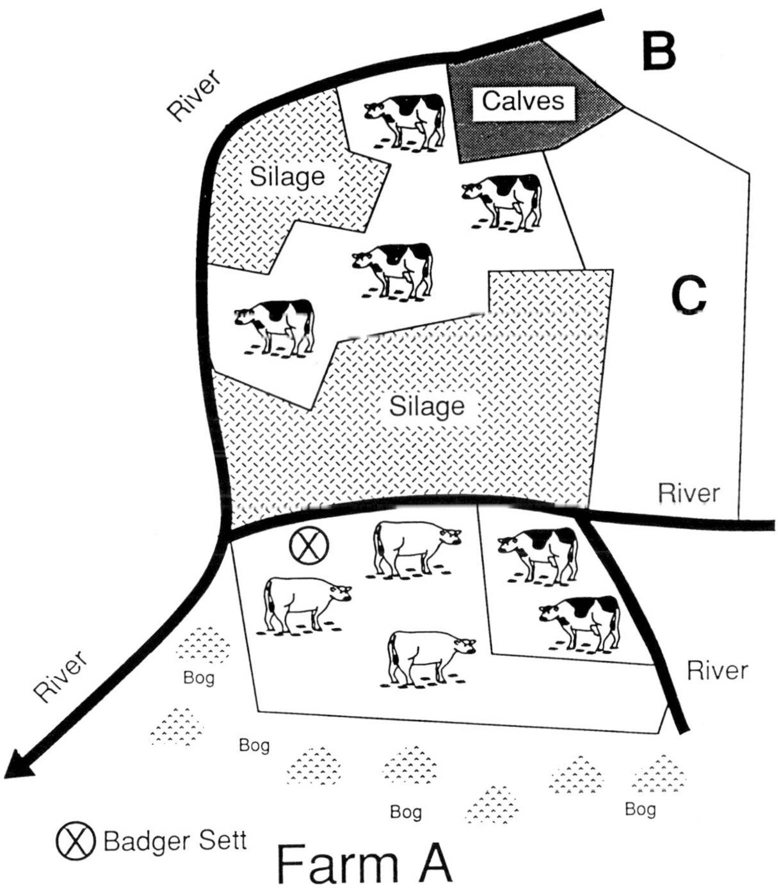

Fig. 2. Disposition of cattle on Farm A and location of adjacent farms B and C. The location of the badger sett is shown at X.

Topographical features

The land is low-lying and flat. For this reason, large drains are excavated on different farms, in which surface water flows towards the Figile River. Such watercourses (which are approximately 10 feet wide x 8 feet deep) form excellent boundaries to the north, west and south-east of Farm A (Fig. 2). Two farms, B and C, adjoin Farm A to the east. There were no cattle on Farm B at the time of this outbreak (sheep only). The herd on Farm C consisted of three or four cattle. This herd had no previous TB history and remained free of TB while Farm A was restricted. Farm A adjoins undeveloped bog to the south (Fig. 2).

Management

Herd A is comprised of three groups: (1) calves, (2) cows, and (3) stores (heifers and bullocks), which are managed separately. The grazing territory of each group is indicated on Fig. 2. The store group grazed in the immediate vicinity of the badger sett. The herd was totally self-contained. The breeding animals were artificially inseminated. Drinking water was piped to troughs in the fields. Slurry was spread on silage ground after the first crop of silage had been harvested. A second crop of silage was subsequently harvested from the same area before any cattle grazed there.

Test details

Prior to 1988, there was no confirmed outbreak of TB in this herd. Details of the reactors disclosed in 1988 and 1989 are summarised in Table 3. During the course of this TB outbreak, 37 reactors were identified in herd A. The majority of the reactors were identified in the store group which grazed immediately beside the badger sett, while only two reactors were identified in the calf group, which grazed at the opposite end of the farm (Fig. 2).

Table 3. Summary of tuberculin status for herd A.

Date	Number tested	Reactors Cows	Reactors Stores	Reactors Calves
3/6/88	134	-	-	-
4/11/88	142	5(4)	16(9)	2(1)
20/1/89	137	5(3)	8(2)	-
31/3/89	154	1(1)	-	-
7/7/89	155	-	-	-

( ) Number with visible TB lesions at slaughter.

Badger activity

At the time of this outbreak, badger activity was evident from the numerous paths, latrines and excavation marks seen throughout the farm. Eight badgers were snared on this farm following the outbreak in November 1988, two of which were confirmed TB-positive. (One had generalised TB, the other had tuberculous liver lesions.)

Conclusions

1. As there were no reactors in any of the herds (even those on the opposite side of the streams) in 1988, lateral spread was not considered a likely source.
2. There was no previous confirmed outbreak of TB in this herd and, as the herd is self-contained, the problem could not be related to residual infection or bought-in animals.
3. Mechanical spread was not considered likely, as all water was supplied in troughs and slurry was spread on silage ground only.
4. There was no evidence to suggest that the infection was introduced by human spread.

Based on the foregoing evidence, it was concluded that the bovine TB problem resulted from the infected badger population on the farm.

DISCUSSION

It has been shown that infectious badgers can transmit TB to cattle under experimental conditions (Little *et al.* 1982). The extent to which transmission occurs under field conditions has not been quantified. The evidence implicating the badger in outbreaks of TB is, for the most part, circumstantial. For example, in Dorset and Thornbury, where experimental badger clearance programmes were undertaken, the incidence of confirmed cases of TB in the cattle population was reduced to nil (Zuckerman 1980). In human medicine there are many areas where 'cause and effect' relationships are accepted on the basis of circumstantial evidence, for example smoking and lung cancer, or thalidomide and its teratogenic effects. In major TB outbreaks, the sighting of a terminally ill/dead badger in close proximity to the infected group(s) of cattle is often reported.

In the two case studies described, the presence of TB in the badger population in the areas was established. In the past, the role of the badger in seeding infection into cattle herds was underestimated mainly because breakdowns resulting from lateral spread closely resemble those caused by wildlife. In recent years, however, it is being increasingly recognised that the badger is particularly important in the transmission of TB to cattle. An analysis of 3975 epidemiology reports from locations throughout Ireland indicated that wildlife was reported as the origin of an outbreak of TB in 14% of cases investigated (Griffin and Hahesy 1992). In the absence of badger control, significant progress in reducing the persistent bovine TB problem in Ireland is not possible (O'Connor and O'Malley 1989). Ireland's TB problem resembles that of south-west England in many respects, and any future strategy must take account of Lord Zuckerman's conclusion regarding that area (SW England):

> 'It seems obvious that no indecisive strategy is going to eliminate the disease in the badger population. Moreover, so long as badgers in the South-west constitute a significant reservoir of the bovine tubercle bacillus, the culling of infected cattle, without a parallel decrease in the population of badgers, is not likely to have much effect on the incidence of herd breakdowns' (Zuckerman 1980).

ACKNOWLEDGEMENTS

I wish to thank Mr Declan Bray, SVI (Tullamore DVO) for his invaluable help and advice in the preparation of this paper, and also to acknowledge the technical help of Professor J.D. Collins UCD and Messrs Leonard Dolan, VI, and John Griffin, VI, Tuberculosis Investigation Unit.

REFERENCES

GRIFFIN, J.M. and HAHESY, T. 1992 Analysis of epidemiology reports on 3975 herd breakdowns in ten DVO regions during 1987-1990. In *Tuberculosis Investigation Unit Selected Papers 1990-1991,* 21. University College, Dublin.

GRIFFIN, J.M., HAHESY, T. and LYNCH, K. 1992 The role of farm management procedures and environmental factors in chronic tuberculosis. In *Tuberculosis Investigation Unit Selected Papers 1990-1991,* 1-6. University College, Dublin.

LITTLE, T.W.A., NAYLOR, P.F. and WILESMITH, J.W. 1982 Laboratory studies of *Mycobacterium bovis* infection in badgers and cattle. *Veterinary Record* **111**, 550-7.

O' CONNOR, R. and O'MALLEY, E. 1989 *Badgers and bovine tuberculosis in Ireland.* Dublin. Economic and Social Research Institute/ERAD.

ZUCKERMAN, LORD 1980 *Badgers, cattle and tuberculosis.* London. HMSO.

In: Hayden, T.J. (ed.) 1993 *The badger*, 139-153. Dublin. Royal Irish Academy.

THE DEVELOPMENT OF FIELD TECHNIQUES FOR STUDYING POTENTIAL MODES OF TRANSMISSION OF BOVINE TUBERCULOSIS FROM BADGERS TO CATTLE

J. A. Brown,* S. Harris*† and C.L. Cheeseman**

* **Department of Zoology, University of Bristol, Woodland Road, Bristol BS8 1UG.**
** **Central Science Laboratory, Ministry of Agriculture, Fisheries and Food, Tangley Place, Worplesdon, Guildford, Surrey GU3 3LQ.**

† To whom correspondence should be addressed.

ABSTRACT

The routes of transmission of tuberculosis from badgers to cattle are still largely speculative, and so novel techniques have been developed to help understand these processes. Spool-and-line tracking in conjunction with a biomarker were used to locate sites of badger urination and defecation. A 0.38g ml^{-1} solution of fluorescein LT injected subcutaneously inhibited *Mycobacterium bovis*, while large numbers of bacilli were recovered with a 0.22g ml^{-1} solution. Thus urine from infected badgers injected with a 0.22g ml^{-1} solution of fluorescein LT was located at night using a portable ultraviolet lamp, so that contaminated grass samples could be removed for subsequent bacterial culture to establish the longevity and vertical distribution of *M. bovis* on pasture. Behavioural interactions between badgers and cattle, and of cattle to contaminated pasture, were also recorded. These techniques will help identify the possible routes of transmission of bovine tuberculosis from badgers to cattle.

INTRODUCTION

Bovine tuberculosis (*Mycobacterium bovis*) was first observed in the European badger (*Meles meles*) in Britain in 1971 following the discovery of a dead badger on a Gloucestershire farm which had recently experienced a herd breakdown. Since that discovery, circumstantial evidence has accumulated linking badgers with the transmission of bovine tuberculosis in the south-west of England (Muirhead *et al.* 1974; Gallagher *et al.* 1976; MAFF 1976; 1977; 1979). This evidence includes the finding that *M. bovis* from infected badgers produces tuberculosis lesions when inoculated into cattle (Little *et al.* 1975), while healthy calves can become infected when in contact with either artificially or naturally infected badgers (Little *et al.* 1982). A close relationship has been found between the incidence of tuberculosis in badgers and herd breakdowns (Muirhead *et al.* 1974). As a result, in 1975 the Ministry of Agriculture, Fisheries and Food introduced a policy of badger control in areas where badgers were believed to be responsible for transmission (Muirhead *et al.* 1974; Gallagher *et al.* 1976).

The route by which bovine tuberculosis might spread from badgers to cattle has been little studied and remains speculative, although it is believed that cattle may pick up infection while grazing pasture contaminated by an infected badger (Muirhead *et al.* 1974; MAFF 1979). This could result from the contamination of pasture, cattle food or both by infected badger urine, sputum and/or faeces (Muirhead *et al.* 1974). Another source of environmental contamination might be pus from infected bite wounds, since *M. bovis* has been recovered from the exudate of wounds in live badgers (Muirhead *et al.* 1974). Although badgers and cattle are generally thought to avoid each other (Benham and Broom 1989), behavioural changes in badgers in an advanced stage of infection have been reported; such badgers appear to lose their fear of man and inhabit farm buildings (Muirhead *et al.* 1974; Cheeseman and Mallinson 1981). Such behaviour may increase the risk of transmission between badgers and cattle, especially if cattle food is being contaminated. Contamination around the sett may also be of significance, since cattle have been observed to rub their heads in the exposed earth around setts and then lick the ground and each other's face (Benham 1984).

Of these various possibilities, badger faeces are likely to be of minor importance in the transmission of *M. bovis* from badgers to cattle because they contain a relatively low number of bacilli, and these are only exposed in small numbers by weathering (MAFF 1976; 1979). Furthermore, faeces are generally deposited in pits in discrete areas (latrines) and a relatively large proportion of these are inaccessible to cattle; cattle also strongly avoid the ingestion of pasture contaminated with faeces (Benham 1985). Conversely, up to 300,000 *M. bovis* ml^{-1} of urine have been recovered from infected badgers (Gallagher and Horwill 1977); with a badger excreting on average about 30ml of urine at a time, this would result in the emission of a considerable number of bacilli. Whilst the pattern of urine deposition by badgers has been poorly documented, it is known that cattle perform more sniffing in response to urine than faeces (Benham 1985). Therefore contaminated badger urine would appear to pose a greater risk than faeces for transmission.

The present study was designed to investigate specific aspects of badger and cattle behaviour in order to identify the factors which contribute to the risk of cross-infection from badgers to cattle, with the ultimate aim of minimising the spread of *M. bovis* infection from badgers to cattle. The main goals for the first part of the study were to develop techniques to study urinary and faecal contamination of farmland, with a greater emphasis on the former, and to identify the type of substrate (e.g. vegetation of a particular height or type) selected by badgers for depositing their excreta. Once the sites of contamination in pasture have been established, samples of grass can be collected and removed for bacterial culture to determine longevity of *M. bovis* on pasture. In addition, the behaviour of badgers and cattle when they are sharing pasture is being studied because their interactions may be important in the transmission process. In this paper we describe the development of the various study techniques and give some results from the preliminary field trials.

THE STUDY AREA

The study area lies in the Cotswold escarpment of Gloucestershire; habitat details are given by Cheeseman *et al.* (1981; 1985). The area supports the highest density of badgers so far recorded, with approximately 20 adults km^{-2} (Cheeseman *et al.* 1981; 1987), and there is a history of herd breakdowns on the study area. These factors, combined with the high prevalence of tuberculosis in badgers in the Cotswold escarpment, mean that this is an ideal site for studying transmission of bovine tuberculosis from badgers to cattle.

Initially we concentrated on one particular social group, which at the onset of the study contained the greatest number of infected badgers in the study area. At the end of 1990, 40% of the group were infected. Details of the group are presented in Table 1, including the progression of tuberculosis infection through the group between January 1988 and November 1990. Only those badgers caught during 1990 are included in the table. The badgers had been marked with a combination of ear tags and tattoos (Cheeseman and Harris 1982); the sampling procedure to establish their tuberculosis status is described by Cheeseman *et al.* (1985; 1988). The group had a territory size of 24.6ha, compared with an average territory size of 23.2ha in the study area (Evans *et al.* 1989).

DEVELOPMENT OF METHODS

Spool-and-line tracking

A technique for tracking badgers was required which would reveal the precise path taken by an animal. Radio-tracking was unsuitable because it can have notoriously poor spatial resolution (Macdonald and Amlaner 1980). Spool-and-line tracking has been used successfully for this purpose on a wide range of mammals, e.g. Miles *et al.* (1981) used it on 263 edentates, marsupials, rodents and carnivores. They attached a spool of thread mid-dorsally on animals smaller than the common opossum (*Didelphis marsupialis*); as the animal moved, it left a trail of thread that could be followed along the route taken. Berry *et al.* (1987) tested the efficacy of spool-and-line tracking on New Guinea giant rats (*Mallomys rothschildi* and *Hyomys goliath*), and Hawkins (1989) applied the technique to badgers; she fitted spool-and-lines to 15 animals and obtained data from seven. The original plan for the construction of the spool-and-line was to develop a device which would allow the line to be released two or three nights after capturing the badger. This would then overcome any potential changes in behaviour following capture and handling. Two devices were tested to control the release of the line. One involved a radio receiver that would release the line on receipt of a signal. However, this proved to be too bulky and battery life was minimal with the equipment involved. Reliance on a radio signal for operation was probably also best avoided. The second design was based on an electronic delayed timing device. The electrical side was relatively simple but problems continually occurred with the mechanical firing mechanism. These devices to delay release of the line were eventually abandoned.

Table 1. Progression of *M. bovis* infection in badgers in the study group.

No.	Sex	Age (mths)	1988 Jan	1988 May	1988 Jul	1988 Oct	1989 Jan	1989 May	1989 Aug	1989 Oct	1990 Jan	1990 Mar	1990 Apr	1990 May	1990 Jul	1990 Sep	1990 Oct	1990 Nov
L61	M	54	-	0	0	-	-	0	-	-	-	-	-	-	F TA D			
K65	F	42	0	-	0	-	-	-	0	-	-	-	-	-	-	0	-	-
K77	F	42	-	0	-	0	-	-	0	-	-	-	-	0	-	-	-	-
K78	F	42	-	0	-	-	-	0	0	-	-	-	0	0	-	-	-	-
E24	M	30	-	0	0	0	0	0	-	0	-	F	-	-	-	-	-	-
E74	M	30	-	-	0	0	0	0	-	-	-	F	-	0	0	-	0	-
E83	M	30	-	-	0	0	0	0	-	0	-	-	-	0	-	-	0	-
A12	M	18	-----	-----	-----	-----	-----	0	0	-	-	F	BW	D				
A13	F	18	-----	-----	-----	-----	-----	0	0	F TA SN	TA SN	TA SN	-	F TA SN	F TA SN U	-	F*	-
A30	F	18	-----	-----	-----	-----	-----	U	0	0	0	-	-	0	-	-	-	-
A74	F	18	-----	-----	-----	-----	-----	-	0	0	-	-	0	0	0	-	0	-
V6	M	06	-----	-----	-----	-----	-----	-----	-----	-----	-----	-	-	0	-	-	-	-
V94	M	06	-----	-----	-----	-----	-----	-----	-----	-----	-----	-	-	-	-	-	0	0

Key: U=*M.bovis* isolated from urine; SN=from open abcess in submandibular lymph node; TA=from tracheal aspirate; BW=from bite wound(s); F=from faeces; 0=*M.bovis* not isolated; D=died;* only faeces collected. All ages given as July 1990.

The spool-and-line device finally adopted was similar to that used by Anderson *et al.* (1988), and did not include any triggering or timing device. Therefore the badger had to be tracked the night following capture. Since this badger population had been trapped extensively as part of a long-term study, it was felt that any stress following capture was likely to be minimal, and so following the animal on the night post-capture was unlikely to give aberrant data. For the initial trials 1500m lines were used, but these were too short to provide a full night's data. Therefore we changed to using 3000m lines; these consisted of bleached white tri-lobal polyester thread (breaking strain 0.9kg) which was enclosed in a polypropylene centrifuge tube (39mm internal diameter) with a 10mm hole at one end allowing the line to run freely. Spools of line were supplied by James Pearsall and Co. Ltd, Tancred Street, Taunton, Somerset TA1 1RY. The spool-and-line device in its centrifuge tube weighed 70g and was fitted parallel to a collar which also carried a radio transmitter (Fig. 1). The centrifuge tube containing the spool-and-line was attached using insulating tape; this would allow the device to fall off within a few days if the animal was not later recaptured. The complete package, including collar, spool-and-line, beta-light and transmitter, weighed 211g and so was well within the guideline of 5% of the body weight suggested by Macdonald (1978). Radio-tracking was undertaken simultaneously with the spool-and-line tracking, with the badger's position recorded at 15min intervals, to provide a time-frame for the spool-and-line data.

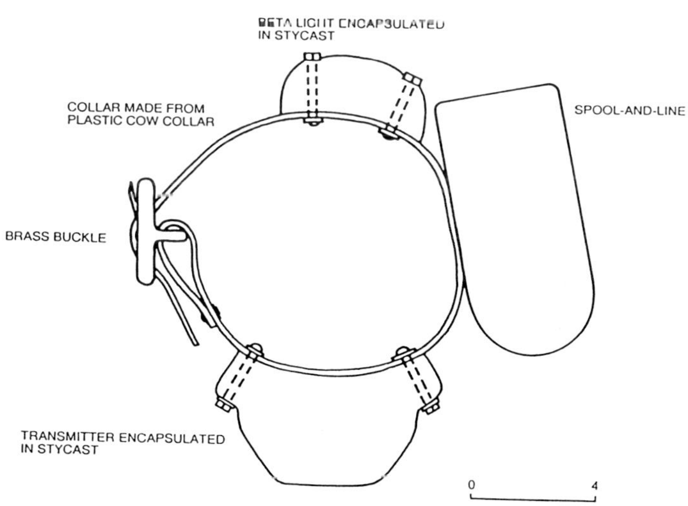

Fig. 1. The arrangement of the spool-and-line on the radio-collar; the spool-and-line is attached with insulating tape and will soon fall off the collar.

Badgers were caught in cage traps (Cheeseman and Mallinson 1980) which were not set until the middle of the night. After being fitted with a spool-and-line, the badger was released in the morning at the hole closest to the capture point and the free end of the thread attached to vegetation in the vicinity of the hole. Care was taken when releasing the badger to ensure that the line was running freely. Although trapping in the middle of the night reduced capture rate, badgers caught this way were usually returning from their feeding grounds with a full stomach. This was important since urination and defecation were likely to proceed as normal the following night. Badgers caught emerging from the sett would not have fed for approximately 36 hours prior to being tracked. This reduced food intake could result in a deviation from normal behaviour and a lack of or reduced rate of urination/defecation during the early part of the night that they are tracked.

Biomarkers to locate badger urine and faeces

A biomarker was needed that would show up in both urine and faeces and could be detected at night, so that samples of contaminated pasture could be removed for culture shortly after the urine and/or faeces had been deposited. Davis *et al.* (1948), New (1958) and Seal and Bhattacharji (1961), in studies of small mammals, used food containing dyes that subsequently showed up in faeces or urine. Frantz (1972) used baits marked with fluorescent pigments and examined the study area at night under ultraviolet illumination for fluorescing rat faeces, tracks and body rub-marks. This technique was considerably more efficient in locating dye traces. Therefore we investigated various fluorescent dyes using a captive colony of badgers, but none proved as easy to locate as fluorescein. Fluorescein has a number of other practical advantages: it is freely soluble in water and produces an intense yellow/green fluorescence perceptible down to 0.04ppm. Also, Webb *et al.* (1962) found that, following oral administration, fluorescein and its mono- and dihalogenated derivatives could be detected in the urine and bile of rats and that the basic fluorescein structure was not degraded in the animals, as shown by the high (90-100%) recovery of fluorescein from excreta. All these findings led us to investigate the use of the dye fluorescein as a biomarker to help locate badger urine and faeces at night using a hand-held portable ultraviolet lamp. The lamp chosen was one that emits ultraviolet light of 366nm wavelength; this will not kill bacilli since ultraviolet light is not bactericidal above 330nm.

For the original trials on captive badgers, fluorescein LT was presented orally in a bait. However, this was non-selective and produced unnecessary contamination with dye in the area of bait release, and so a different technique was needed for field use. Two captive badgers were used to compare subcutaneous injection and oral administration of a 0.38g ml^{-1} solution of fluorescein LT. For the oral administration a dog catheter (length 50cm, outer diameter 2.8mm) was used to inject the dye directly into the stomach of a 12-year-old sow; a 2-year-old sow was used for the subcutaneous injection. The two badgers were then held in separate cages (1.2m x 0.9m x 1.1m) and supplied with bedding, water and dead day-old chicks *ad libitum*. Each cage was raised off the ground and a polythene sheet placed underneath. Pockets were formed in the sheet using broom handles so that successive urinations were less likely to mix. The sheet was examined at three-hourly intervals for the following five days and any urine and faeces collected. The quantity of urine produced at each urination varied considerably, ranging from 5.5ml to

85ml. The presence of fluorescein in the urine, faeces and gelatinous anal gland secretion was revealed at night using the ultraviolet lamp.

Following subcutaneous injection, fluorescein was detectable in the urine for over 93hrs after injection and more than 144hrs for the faeces; the badger was still excreting fluorescein in its faeces after it had been returned to the captive colony. By comparison, the catheterised badger produced the first negative urine after 63hrs and no more fluorescein was detected in the urine after 84hrs. Faeces of this animal no longer contained the dye after 45hrs. Clearly subcutaneous injection resulted in the longest retention of the fluorescein, and was subsequently used for the field trials.

Effect of the biomarker on M. bovis

Trials were undertaken to examine the effect of a 0.38g ml^{-1} solution of fluorescein LT on the growth of *M. bovis*. A dilution series (10^{-1} to 10^{-4}) of *M. bovis* was prepared in saline to give low numbers of organisms present in the dilutions. A sample (0.1ml) of each dilution was added to 0.9ml of each of (i) saline, (ii) urine and dye, (iii) urine alone. From each treatment, 0.25ml was sown onto each of three 7 H 11 slopes. This initial trial (Table 2) showed that a 0.38g ml^{-1} solution of fluorescein LT appeared to inhibit the growth of *M. bovis*.

Table 2. Effect of a 0.38g ml^{-1} solution of the dye fluorescein LT on the isolation of *M. bovis* from badger urine (low dose of *M. bovis* used).

Sample	**Dilution**	**Replicate**		
		1	**2**	**3**
Saline	10^{-1}	12b/c	4b/c	14b/c
	10^{-2}	-	3b/c	-
	10^{-3}	-	-	-
	10^{-4}	-	-	-
Urine + dye	10^{-1}	-	-	-
	10^{-2}	-	-	-
	10^{-3}	-	-	-
	10^{-4}	-	-	-
Urine only	10^{-1}	5b/c +1col	3b/c	3b/c
	10^{-2}	-	-	1b/c
	10^{-3}	-	-	-
	10^{-4}	-	-	-

Key: b/c = 'breadcrumb' (colonies in condensation water); col = colony on slope; - = no colonies present.

To establish the degree of inhibition of *M. bovis* by a 0.38g ml^{-1} solution of fluorescein LT, the experiment was repeated using a higher dose of *M. bovis* with dilutions ranging from 10^{-1} to 10^{-6}. Colonies were present on the slopes of the saline treatment down to a dilution of 10^{-5} and 10^{-4} for the urine treatment, although, as with the previous experiment, *M. bovis* could not be isolated in the presence of the dye (Table 3). It was possible that the concentration of fluorescein present in the urine was critical and so a range of different dye concentrations were examined. Table 4 shows that considerable numbers of bacilli were recovered when a 0.22g ml^{-1} solution of fluorescein LT was used, and that more bacilli were recovered using fluorescein LT compared to fluorescein LTS. The former is a technical grade dye that contains fewer impurities; these results suggest that a 0.22g ml^{-1} solution of fluorescein LT is suitable for field use.

Table 3. Effect of a 0.38g ml^{-1} solution of the dye fluorescein LT on the isolation of *M. bovis* from badger urine (high dose of *M. bovis* used).

Sample	Dilution	Replicate		
Saline	10^{-1}	+++	+++	+++
	10^{-2}	+++	+++	+++
	10^{-3}	++	++	++
	10^{-4}	b/c	b/c	-
	10^{-5}	+	-	-
	10^{-6}	-	-	-
Urine+dye	10^{-1}	-	-	-
	10^{-2}	-	-	-
Urine only	10^{-1}	+++	+++	+++
	10^{-2}	+++	++	++
	10^{-3}	+	b/c	b/c
	10^{-4}	+	+	+
	10^{-5}	b/c	-	-
	10^{-6}	-	-	-

Key: +++ = >50 colonies on slope; ++ = 10-50 colonies; + = <10 colonies; b/c = 'breadcrumb' (colonies in condensation water); - = no colonies present.

Spatial and temporal distribution of badgers and cattle

Two fields situated close to the main sett were divided into a grid system with posts at 50m intervals. The squares formed by the posts were subdivided visually into quarters so that badgers and cattle could be positioned in cells of 25m x 25m.Reflective tape was nailed on to the top of the posts, thereby facilitating their location at night with image intensifiers and an infra-red light source. The fields were observed from a 6m-high observation tower situated between the two fields. The tower facilitated viewing of the fields as well as reducing the possibility of detection by the badgers. The field was scan-sampled at 5min intervals; the positions of the badgers and cattle were recorded, as was the behaviour of the cattle. When a badger injected with dye entered the field its position was recorded continuously so that the time and location of each urination could be established.

Table 4. Growth of *M. bovis* in badger urine and with different concentrations and types of fluorescein three weeks after inoculation.

Dilution	Saline	Urine only			Urine+dye				
					Fluorescein LT		Fluorescein LTS		
					0.12g ml^{-1}	0.22g ml^{-1}	0.12g ml^{-1}	0.22g ml^{-1}	0.38g ml^{-1}
10^{-1}	+++	+++	+++	+++	+++	+++	-	+++	+++
	+++	++	+	+++	-	+++	-	+++	-
	+++	+	-	+++	-	+++	-	+++	-
10^{-2}	+++	+	+++	+++	+++	+++	+++	+++	-
	+++	+	+++	+++	+	+	+++	-	-
	+++	-	+	+++	-	+	-	-	-
10^{-3}	+	+	++	++	++	++	++	+	+
	+	-	+	++	++	++	-	-	-
	-	-	+	++	-	+	-	-	-
10^{-4}	++	+	-	+	+	+	-	+	-
	+	+	-	+	-	-	-	-	-
	-	-	-	I	-	-	-	-	-
10^{-5}	-	+	-	+	+	-	-	+	-
	-	-	-	+	+	-	-	-	-
	-	-	-	+	+	-	-	-	-
10^{-6}	+	+	+	+	-	-	-	+	-
	-	+	-	+	-	-	-	-	-
	-	-	-	+	-	-	-	-	-

Key: +++ = >100 colonies on slope; ++ = 20-100 colonies; + = 1-20 colonies; - = no colonies isolated.

RESULTS FROM THE PRELIMINARY FIELD TRIALS

Spool-and-line tracking

The first three 3000m lines were attached to different badgers in April and May. Each produced a complete night's data without a break in the line; the lengths of line produced were 1333m, 1591m and 1901m. Distance measurements were calculated from a known length : mass ratio for the thread. One example of a night's activity is shown in Fig. 2. Five urination and two defecation sites were located over the 1901m of thread searched. The first four urinations occurred at latrines, with the fifth situated beside a run. Four of the five sites were of easy access to cattle. All five urinations were in the form of a trail of urine, rather than a discrete patch, suggesting movement of the badger while urinating. The two defecations occurred at latrines, with one accompanied by anal gland secretion.

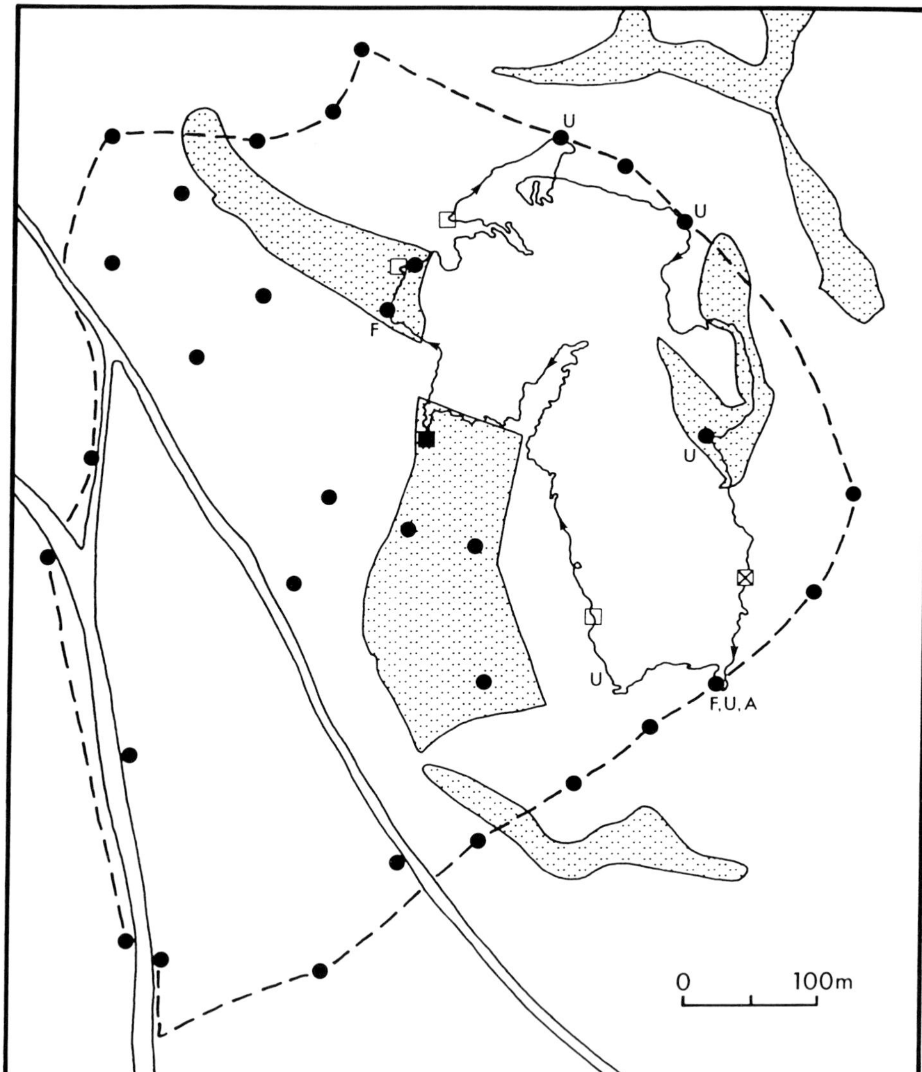

Fig. 2. The results from one night's spool-and-line tracking of an adult female badger; 1901m of line was followed from the point at which the badger emerged from a main sett (solid square). The outer line demarks the territorial boundary, circles=latrines, dotted areas=woodland, open squares=outlying setts: the badger briefly entered the area marked with a cross. U=urination point, F= defecation point, A=anal gland secretion.

Interactions between badgers and cattle

For the first season, the cattle held on the study field were a group of four breeding cows 4-8 years old. *T*-tests were used to compare the observed distribution of each badger's path in the field compared to a random path; these revealed no significant differences. Clearly the badgers were not limiting their activity to those parts of the field not occupied by the cows. The cows would spend most of each night sitting down, and foraging badgers would approach within 5m of sitting cows. The badgers appeared to take no notice of the cows, and cows were rarely observed even to watch the movements of the badgers. The response of badgers and cattle that have regular exposure to each other, as in the present study, appears from the initial results to be one of mutual indifference. Whether this behaviour is significant in the transmission process will be determined by more detailed observations.

DISCUSSION

These techniques show great potential for helping to understand the routes of transmission of TB from badgers to cattle. Spool-and-line tracking can show the precise route taken by an individual animal and when used in conjunction with a biomarker it allows the location of the urination and defecation sites of specific animals. There are, however, a number of problems associated with this particular type of tracking. According to Berry *et al.* (1987), the pattern of thread deposition provides an indication of the time spent in each area, i.e. stationary (concentrated jumble), moving slowly (slack loops) or travelling fast (taut). However, the pattern of thread deposition also varies according to the particular habitat type, as different types of vegetation catch the thread at different frequencies. Therefore, in the present study radio-tracking was used in conjunction with the spool-and-line technique to provide a time element for the analysis of the animal's movements.

Other problems are more difficult to resolve. Grazing animals can cause problems by dragging the thread from its original position with their feet and also by eating it. Animals which forage on pasture, such as foxes (*Vulpes vulpes*) and hedgehogs (*Erinaceus europaeus*), as well as other badgers, can also alter the position of the thread. Wind can distort the position of the thread as it is being released and also when on the ground, especially in open fields where there are relatively few points of attachment. These problems are minimised by following the thread at the end of the night, before its position has been significantly altered.

Fluorescein has been extensively tested on a variety of micro-organisms for mutation and DNA alteration. The DNA-cell-binding test suggested that this dye was a possible mutagen on metabolic activation (Kubinski *et al.* 1981). Nestmann *et al.* (1979) and Douglas *et al.* (1980) reported rhodamine B to be mutagenic in *Salmonella* and mammalian cells and demonstrated that the mutagen was an impurity in the technical grade dye employed. Their result has important implications, because the concentration of impurities in commercial dyes may vary with both manufacturer and dye batch, and so each batch of dyes should be tested before it is used to locate urine patches from which bacilli are to be cultured.

There is little quantified information about the pattern and frequency of urine deposition on badger territories. Paget and Middleton (1974) observed badgers urinating in special latrine pits (i.e. pits not used for faeces or anal gland secretion). Kruuk (1978) reported badgers 'squat-marking', during which they released glandular secretions as well as urine, both on and within the territory boundary. Eunice Overend (personal communication) suggests that urination can be performed in two different contexts. She has seen urine deposited in small hollows in the ground as described above, and also suggests that boars may mark prominent herbage with urine soon after leaving the sett. Many of the patches of urine located in this study were trails, with some measuring over a metre in length, indicating that the badgers were moving during urination. A greater area of contamination would result from a trail as opposed to a discrete patch of urine, and these trails may be less easily avoided by grazing cattle, especially the spots of urine often found at the end of these trails. Further field trials observing the behaviour of cattle in relation to urine-contaminated pasture will show whether this is in fact the case.

The overall pattern of faecal deposition on badger territories is better understood, although the defecation behaviour of individual animals is not. In addition to the normal latrines, a second type of defecation site has been identified (termed a temporary defecation site) that is probably not concerned with territorial defence (Roper *et al.* 1986). These temporary defecation sites are often located away from the edge of a field, are generally smaller than a latrine and are used only once or twice before being abandoned. For these reasons they could possibly be of greater significance in terms of transmission of TB to cattle than latrines, and spool-and-line tracking will be particularly useful for locating such sites, monitoring the survival of bacilli, and observing the behaviour of grazing cattle in relation to these temporary defecation sites.

The methodologies described here have enabled us to study various aspects of badger behaviour hitherto little understood. By collecting further data on the pattern of pasture contamination by infected badgers, the survival of *M. bovis* on pasture, and the behaviour of cattle towards both badgers and contaminated pasture, a better understanding of the potential routes of transmission may emerge. With this increased understanding it may be possible to design measures to reduce the risk of transmission of bovine tuberculosis from badgers to cattle.

ACKNOWLEDGEMENTS

The considerable cooperation received from the farming community and the expert assistance of the Ministry of Agriculture, Fisheries and Food's field staff and the technical staff at the Central Veterinary Laboratory, particularly Paul Croston, are gratefully acknowledged. JAB thanks the Natural Environment Research Council for a studentship, and the Ministry of Agriculture, Fisheries and Food and the Universities Federation for Animal Welfare provided additional funds to purchase equipment for the project.

REFERENCES

ANDERSON, T.J.C., BERRY, A.J., AMOS, J.N. and COOK, J.M. 1988 Spool-and-line tracking of the New Guinea spiny bandicoot, *Echymipera kalubu* (Marsupialia, Peramelidae). *Journal of Mammalogy* **69**, 114-20.

BENHAM, P.F.J. 1984 Social organisation in groups of cattle and the interrelationships between social and grazing behaviours under different management systems. Unpublished Ph.D. thesis, University of Reading.

BENHAM, P.F.J. 1985 A study of cattle and badger behaviour and farm husbandry practices relevant to the transmission of bovine tuberculosis *(Mycobacterium bovis)*. Unpublished project report submitted to the Ministry of Agriculture, Fisheries and Food.

BENHAM, P.F.J. and BROOM, D.M. 1989 Interactions between cattle and badgers at pasture with reference to bovine tuberculosis transmission. *British Veterinary Journal* **145**, 226-41.

BERRY, A.J., ANDERSON, T.J.C., AMOS, J.N. and COOK, J.M. 1987 Spool-and-line tracking of giant rats in New Guinea. *Journal of Zoology* **213**, 299-303.

CHEESEMAN, C.L. and HARRIS, S. 1982 Methods of marking badgers (*Meles meles*). *Journal of Zoology* **197**, 289-92.

CHEESEMAN, C.L. and MALLINSON, P.J. 1980 Radio tracking in the study of bovine tuberculosis in badgers. In C.I. Amlaner and D.W. Macdonald (eds), *A handbook on biotelemetry and radio tracking*, 649-56. Oxford. Pergamon Press.

CHEESEMAN, C.L. and MALLINSON, P.J. 1981 Behaviour of badgers (*Meles meles*) infected with bovine tuberculosis. *Journal of Zoology* **194**, 284-9.

CHEESEMAN, C.L., JONES, G.W., GALLAGHER, J. and MALLINSON, P.J. 1981 The population structure, density and prevalence of tuberculosis (*Mycobacterium bovis*) in badgers (*Meles meles*) from four areas in south-west England. *Journal of Applied Ecology* **18**, 795-804.

CHEESEMAN, C.L., LITTLE, T.W.A., MALLINSON, P.J., REES, W.A. and WILESMITH, J.W. 1985 The progression of bovine tuberculosis infection in a population of *Meles meles* in south-west England. *Acta Zoologica Fennica* **173,** 197-99.

CHEESEMAN, C.L., WILESMITH, J.W., RYAN, J. and MALLINSON, P.J. 1987 Badger population dynamics in a high-density area. *Symposium of the Zoological Society of London* **58**, 279-94.

CHEESEMAN, C.L., WILESMITH, J.W., STUART, F.A. and MALLINSON, P.J. 1988 Dynamics of tuberculosis in a naturally infected badger population. *Mammal Review* **18**, 61-72.

DAVIS, D.E., EMLEN, J.T. and STOKES, A.W. 1948 Studies on home range in the brown rat. *Journal of Mammalogy* **29**, 207-25.

DOUGLAS, G.R., GRANT, C.E., BELL, R.D.L., WYTSMA, J.M., NESTMANN, E.R. and KOWBEL, D.J. 1980 Effect of impurities and metabolic activation on the mutagenic activity of rhodamine B in bacterial and mammalian cells. *Environmental Mutagenesis* **2**, 289-90.

EVANS, P.G.H., MACDONALD, D.W. and CHEESEMAN, C.L. 1989 Social structure of the Eurasian badger (*Meles meles*): genetic evidence. *Journal of Zoology* **218**, 587-95.

FRANTZ, S.C. 1972 Fluorescent pigments for studying movements and home ranges of small mammals. *Journal of Mammalogy* **53**, 218-23.

GALLAGHER, J. and HORWILL, D.M. 1977 A selective oleic acid albumin agar medium for the cultivation of *Mycobacterium bovis*. *Journal of Hygiene* **79**, 155-60.

GALLAGHER, J., MUIRHEAD, R.H. and BURN, K.J. 1976 Tuberculosis in wild badgers (*Meles meles*) in Gloucestershire: pathology. *Veterinary Record* **98**, 9-14.

HAWKINS, C. 1989 A preliminary methodological investigation of the feasibility and potential of spool-and-line tracking the European badger (*Meles meles*). Unpublished B.Sc. project, St Catherine's College, University of Oxford.

KRUUK, H. 1978 Spatial organization and territorial behaviour of the European badger (*Meles meles*). *Journal of Zoology* **184**, 1-19.

KUBINSKI, H., GUTZKE, G.E. and KUBINSKI, Z.O. 1981 DNA-cell-binding (DCB) assay for suspected carcinogens and mutagens. *Mutation Research* **89**, 95-136.

LITTLE, T.W.A., BURN, K.J. and STUART, P. 1975 Virulence for calves of tubercle bacilli isolated from badgers. *Veterinary Record* **96**, 533.

LITTLE, T.W.A., NAYLOR, P.F. and WILESMITH, J.W. 1982 Laboratory study of *Mycobacterium bovis* infection in badgers and calves. *Veterinary Record* **111**, 550-7.

MACDONALD, D.W. 1978 Radio-tracking: some applications and limitations. In B. Stonehouse (ed.), *Animal marking: recognition marking of animals in research*, 192-204. London. Macmillan Press.

MACDONALD, D.W. and AMLANER, C.I. 1980 A practical guide to radio tracking. In C.I. Amlaner and D.W. Macdonald (eds), *A handbook on biotelemetry and radio tracking*, 143-59. Oxford. Pergamon Press.

MAFF 1976 *Bovine tuberculosis in badgers*. London. Ministry of Agriculture, Fisheries and Food.

MAFF 1977 *Bovine tuberculosis in badgers*. Second report. London. Ministry of Agriculture, Fisheries and Food.

MAFF 1979 *Bovine tuberculosis in badgers*. Third report. London. Ministry of Agriculture, Fisheries and Food.

MILES, M.A., DE SOUZA, A.A. and PÓVOA, M.M. 1981 Mammal tracking and nest location in Brazilian forest with an improved spool-and-line device. *Journal of Zoology* **195**, 331-47.

MUIRHEAD, R.H., GALLAGHER, J. and BURN, K.J. 1974 Tuberculosis in wild badgers in Gloucestershire: epidemiology. *Veterinary Record* **95**, 552-5.

NESTMANN, E.R., DOUGLAS, G.R., MATULA, T.I., GRANT, C.E. and KOWBEL, D.J. 1979 Mutagenic activity of rhodamine dyes and their impurities as detected by mutation induction in *Salmonella* and DNA damage in Chinese hamster ovary cells. *Cancer Research* **39**, 4412-17.

NEW, J.G. 1958 Dyes for studying the movements of small mammals. *Journal of Mammalogy* **39**, 416-29.

PAGET, R.J. and MIDDLETON, A.L.V. 1974 Some observations on the sexual activities of badgers (*Meles meles*) in Yorkshire in the months December to April. *Journal of Zoology* **173**, 256-60.

ROPER, T.J., SHEPHERDSON, D.J. and DAVIES, J.M. 1986 Scent marking with faeces and anal secretion in the European badger (*Meles meles*): seasonal and spatial characteristics of latrine use in relation to territoriality. *Behaviour* **97**, 94-117.

SEAL, S.C. and BHATTACHARJI, L.M. 1961 Epidemiological studies on plague in Calcutta. II. The role of movement of rats in the spread of plague infection (with special reference to the epidemic plague in Calcutta, 1948-51). *Indian Journal of Medical Research* **49**, 1008-18.

WEBB, J.M., FONDA, M. and BROUWER, E.A. 1962 Metabolism and excretion patterns of fluorescein and certain halogenated fluorescein dyes in rats. *Journal of Pharmacology and Experimental Therapeutics* **137**, 141-7.

In: Hayden, T.J. (ed.) 1993 *The badger*, 154-165. Dublin. Royal Irish Academy.

BEHAVIOUR OF IRISH BADGERS IN RELATION TO BOVINE TUBERCULOSIS

D.P. Sleeman† and M.F. Mulcahy

Zoology Department, University College, Cork

†To whom correspondence should be addressed.

ABSTRACT

Bovine tuberculosis is widespread in Irish badgers, and it has been argued that they act as a reservoir of the disease although the route(s) by which tuberculosis from badgers might infect cattle is unclear. There is little information on interactions between badgers and cattle. In Ireland a badger population in County Cork which contains infected individuals has been studied for four years. The setts have been mapped, and badgers are caught, sampled, marked and radio-tagged. Radio-tracking has been used to study badger movements, interactions with cattle and mortality. The badgers normally avoided cattle; however, visits to cattle facilities have been recorded. Despite the presence of some infected badgers in the area, no confirmed cases of tuberculosis in cattle have been recorded since 1987.

INTRODUCTION

Bovine tuberculosis is re-emerging as a threat to the health of both man and animals (Collins and Grange 1987; Morris and Pfeiffer 1991). The presence of bovine tuberculosis in Irish badgers (*Meles meles*), and the perceived threat by them to the eradication of the disease in cattle, currently presents one of our most complex and important biological problems. Understanding this problem involves knowledge of bovine tuberculosis epidemiology, badger/cattle interactions and agricultural economics. The implications reach into conservation, ethics and the national economy. One of the first reviews to mention the possible role of the badger gave it a low priority (O'Connor 1986). Tuberculosis is now widespread in Irish badgers, occurring in 31 of the 32 counties, County Armagh being the exception (Dolan 1992; Department of Agriculture (NI), unpublished). This contrasts with 18 counties in Britain recorded with badgers affected by tuberculosis (Cheeseman *et al.* 1989; Cheeseman, pers. comm.). Circumstantial evidence of tuberculosis being transmitted from badgers to cattle has been reviewed by O'Connor and O'Malley (1989), Downey (1990), and Flanagan (1993). Others have argued that such evidence is inadequate (Barry 1990).

The badger project at University College Cork has been in progress since 1987. Its aims are to monitor the movements and behaviour of both cattle and badger populations and to identify risk situations where transfer of infection may occur. It is hoped that by identifying and understanding such situations, strategies may be designed to avoid them. This paper is an interim account of methodology and preliminary observations for the period autumn 1987 to spring 1991.

STUDY AREA AND METHODS

The initial study area was approximately 12km^2 in rural County Cork, but was expanded by 2km^2 in 1990 to include two new setts. It is mainly pastureland but there is also mixed woodland, including forestry holdings, streams and lakes. There are fifteen cattle herds and seventeen major landowners. Badgers have been in one sett (sett A) in the area for at least 90 years (Warren 1892); however, the area as a whole has a history of intentional and accidental persecution by badger-diggers, fox-trappers and fox-hunters. The area had a history of unexplained tuberculosis breakdowns in cattle herds prior to the study.

Badgers are captured, under licence, by trapping and snaring. Both techniques are based on the recommendations of Cheeseman and Mallinson (1979). However, a number of changes have been made in the details of these methods. Traps have been modified and now have a smaller (2.54cm) mesh to prevent self-inflicted damage by captured badgers attempting to escape. Sacking is placed on the floor of the traps to prevent badgers acquiring the peanut bait without entering the traps. To prevent unauthorised use, the trap mechanisms can now be removed when not in use. Badgers are restrained while being anaesthetised. This is done using wooden rods pushed into the mesh traps; for snared badgers, a forked stick and twisted steel rod is used. Ketamine hydrochloride ('Vetalar') is used as an anaesthetic, injected intramuscularly at a dose of approximately 20mg/kg body weight. This anaesthetic has had widespread successful application for badgers (e.g. Cheeseman and Mallinson 1979; Parish and Kruuk 1983). Snaring has not yet been widely applied. However, it will be used as the routine capture technique for adult badgers because of ease of use and effectiveness. It is not suitable for cubs in the spring and summer. Untagged adult animals appear from time to time, and the extent to which these are migrants or undetected resident badgers is not known.

Captured badgers are weighed and ear-tagged (Dalton 'Rototags'), and blood, faecal, urine and swab samples may be taken. Adults are radio-tagged (see below). In addition to faecal samples from live badgers (n=10), faecal samples (n=111) were obtained from latrines at regular intervals. Urine samples (n=5) were also taken. Faecal and other samples have been tested by culture in the bacteriology sections of Cork Regional Hospital and the Veterinary Research Centre, Abbotstown. Blood samples (n=15) were taken and stored as frozen plasma. It is intended that, with the development of an efficient blood test for tuberculosis, these samples will be assayed to reveal which badgers were infected. In cooperation with the MAFF Central Veterinary Laboratory, Weybridge, some duplicate blood samples have already been tested, using a monoclonal blocking antibody test (Nolan and Goodger 1994). Badgers found dead in the study area are subjected to examination post-mortem at the Regional Veterinary Laboratory, Cork.

Social group range and boundaries are established by feeding plastic pellets mixed with peanuts and treacle. The mixture is fed for two weeks, or more, at a rate of approximately 8kg per sett per week. Differently coloured plastic pellets are fed at main setts in spring, and latrines at the edge of the social group range (referred to as frontier latrines) are searched for such pellets. This technique has been widely applied (Kruuk 1978; Cheeseman and Mallinson 1979; O'Corry Crowe *et al.* 1993). Badger setts are categorised as 'main', 'annexe', 'subsidiary' or 'outlier', using the definitions of Thornton (1988). The setts are referred to here as setts A through K. Two of the setts are known to contain infected badgers.

Twelve badgers have been fitted with radio-tags (Biotrack, Wareham). Each tag contains a miniaturised transmitter, which allows the badger to be identified, located (radio-location) and followed (radio-tracking). The tags worked for a limited period. Tags which had ceased to function were removed where possible (Table 1).

Table 1. Details of badgers radio-tagged for location and tracking.

TAG NO.	SEX	SOCIAL GROUP	RADIO-TAG		NO. OF NIGHTS TRACKED
			ON	OFF	
*5	F	A	April 1989	Aug 1989	4
*7	M	A	May 1989	June 1990	4
14	F	C	May 1989		1
15	M	E	May 1989	June 1990	2
17	F	E	May 1989	May 1990	4
18	F	D	August 1989	November 1990	0
25	F	C	August 1989		0
26	M	D	December 1989		0
27	F	F	February 1989		0
28	F	F	November 1989	August 1990	0
29	M	F	February 1989		0
30	F	F	May1990	August 1990	8

*Nos 5 and 7 are the only badgers which have proved positive for TB to date.

The radio-tags use a two-stage transmitter with an internal integral loop antenna and magnet-operated switch. They transmit at 151MHz, the usual frequency used in North America for radio-tagging. The frequency is more suitable underground than 173MHz, which is used in Britain. It is not known if transmitters on badgers in setts were always detected and it is possible that in deeper parts of the sett they were out of range. However, experience in recovering detached tags both in this study and elsewhere (Roper and Christian 1992) suggests that at depths of 1-2m they are readily detected. Therefore absence from a sett, at the two main setts regularly monitored, was recorded as such only if, during the period of an absence, the badger was located elsewhere.

The tags are attached by a whole chrome tanned leather collar, which is both supple and durable and is secured by bolted metal plates (Kenward 1987). Beta-lights (Sanders-Roe, Middlesex) encapsulated in perspex and bearing the contact address of the project are fitted to the back of each collar. These contain phosphorescent tritium and can be seen in darkness from 100-200m away, which assists in the exact location of badgers during radio-tracking. Collars of this weight (approx 200g) and fitting cause minimum discomfort to the badgers (Cheeseman and Mallinson 1979; Harris 1982). A three-element aerial (Yagi) and M 57 receiver (Mariner Radar, Lowestoft) are used to receive signals. Night vision equipment (infra-red night scope: Kay Optical, London; and Spylux image intensifier: Davin Optical, Barnet) are also used depending on circumstances. A total of 23 nights have been spent radio-tracking up to spring 1991. Normally one badger was tracked; however, other badgers, both tagged and untagged, were encountered a total of five times during tracking. The badgers are located as they emerge from their setts, and are followed on foot. The badgers are tolerant of the tracker as long as they do not smell him/her. Once a badger has been frightened, it will usually retreat to the main sett. There have been some problems with radio-tags involving transmitter failure and lack of anticipated range.

Cattle numbers are recorded in the field by interviewing herd owners. Cattle testing results are obtained from Cork District Veterinary Office. For details of the cattle tests see Monaghan *et al.* (in press).

RESULTS

The badger population

Trapping success varied seasonally. Best results were achieved in spring, with summer being the least successful season. In summer 1990, for example, one badger was captured for every 12 trap-nights, in spring one per 3 trap-nights, in autumn one per 8 trap-nights and in winter one per 6.

Between 1987 and 1991, a total of 41 individual badgers were identified in the area; most (35) were caught, but six (five cubs and one adult) were found dead without ear-tags. Twelve badgers have been found dead. At the end of the study it was estimated that there are 29 badgers living in the area. The density of badgers, though variable over time and space, is approximately 2.5-3 per km^2. Female badgers outnumber males at a ratio of 1.6:1.

The badger mortalities which have been detected were mainly due to road accidents (8 out of 12) and were mainly cubs and yearlings (10 out of 12). Two cubs in an advanced state of decomposition were found at setts in summer, and two yearlings in a similar state were found on pasture in spring. It is not known if these were infectious. However, they were within the territory of social group B, which contained infectious badgers (see below).

In total, 25 setts have been identified: eleven main setts, three annexe, nine subsidiary and two outliers (Fig. 1). Cattle have access, from at least one direction, to nine of the eleven main setts. Radio-location results indicate that badgers are usually found in the main sett and that the main setts are roughly at the centre of the social group range.

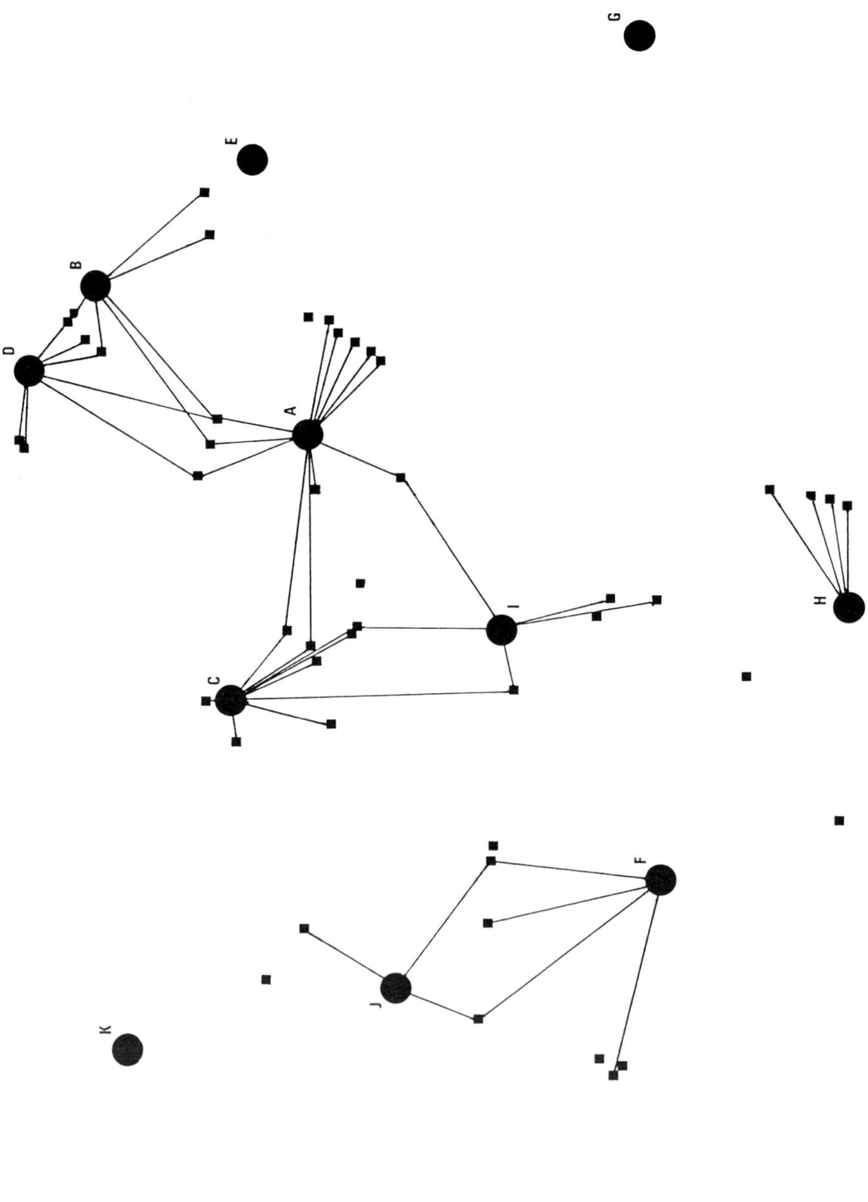

Fig. 1. Relative positions of setts and latrines in the study area. Latrines containing pellets are shown connected to the sett of origin. Main setts • and latrines ▪. Joined by lines as appropriate to show social group ranges in the study area for 1990 (from Sleeman 1992).

Some main setts have become occupied (or re-occupied) during the study and two were deserted. In one instance the badgers (from sett E) deserted their main sett and moved to another, adjacent and apparently deserted main sett, establishing a different social group range as well (Sleeman 1992).

The numbers of badgers per social group ranges from 1 to 11. There were 7 social groups, in 9 main setts, in 1989 (mean size 3.1 badgers), and 9 social groups, in 11 main setts, in 1990 (mean group size 3.5 badgers). There was little observed territoriality in areas where there are only one or two badgers. The territories, as detected by plastic pellets recovered from latrines, tend to be easily defined only in the larger social groups, and are usually more than 1km in width (Sleeman 1992). Many frontier latrines have been located on, or near, the same sites from year to year. Badgers have been protected at sett A by the landowners. This sett had the most individuals: 7 badgers in 1987, 10 in 1989, and 11 in 1990. There was a slight increase overall in the badger numbers in the study area (see above). However, persecution in the area continues; one sett showed signs of being dug in the summer of 1990. One sett was bulldozed in 1988, and recently (March 1990) slurry was pumped onto two setts on one farm.

The cattle population

Overall cattle numbers are approximately 1,500 in fifteen herds. The largest herd numbered 283. The remainder of the herds varied in number from 4 to 170. These figures are not static, as stock movements and mortality, most frequently among calves, caused the overall numbers to vary. The composition of each herd also varied, some being milk herds, some suckler, some beef and some a mixture of these. There is little 'buying in'. However, cattle are brought into the area, from out-farms belonging to people who live in the study area, and to grazing let to farmers living outside the area. Some cattle are out on pasture all year round, but the majority are housed in winter, going out to grass typically in March or April and being housed again in November, December or exceptionally January. Intensive monitoring of summer grazing patterns of cattle in one area has shown that dairy cattle get the first opportunity on fresh grass. Mixtures of other cattle (heifers, calves, bulls etc.) tend to be grazed later on the same fields to 'clean up'.

Bovine tuberculosis

Bovine tuberculosis was detected on three occasions in badgers during the study. Two of these were badgers from sett A (nos 5 and 7). One, a female (no. 5), was killed by road traffic in 1989 and tuberculosis was detected post-mortem. The second case was a male (no. 7) whose tuberculosis was detected by blood test, from a sample taken in 1990. Both these badgers had been radio-tracked. The third detection was in badger faeces from a latrine at sett B. It is safe to assume that both of the social groups involved, which share frontier latrine pits in common, were infected during 1989/90, and possibly were infected for longer. However, given that the most reliable techniques for detecting tuberculosis in badgers are available only post-mortem, and given the poor quality of most of the currently available sampling methods, it is likely that some of the other badgers are infected and as yet remain undetected. Thus an accurate measure of the prevalence of the disease in badgers is not possible based on the data available, but we do know that at least part of the population is infected and sometimes infectious (i.e.

infectious faeces).

All cattle on twelve of the fifteen major holdings were tested as clear of tuberculosis during the four years (1987-90) of the study. On the other three holdings there were a total of fourteen reactors, only one of which (in 1987) had visible lesions on examination post-mortem. Inconclusive results were obtained for 27 animals on these three holdings. Some of these were animals sent voluntarily by their owners to be slaughtered and none had visible lesions. Finally, five common rats (*Rattus norvegicus*) were shot at sett A in 1989, principally because they were eating the peanut bait. Despite the fact that infected badgers were present, and that at least some of the rats lived in the sett, no visible lesions were found at post-mortem.

Badger behaviour

Nine badger encounters with cattle were recorded during 23 nights of radio-tracking. On these occasions the badgers always avoided the cattle. Furthermore, badgers did not emerge if cattle were present at the entrances to the setts. On the other hand, cattle were noted to move closer to badgers, and on one occasion (in July 1989) actually chased a badger they encountered. Cattle regularly visited the entrances of three main setts and have caused tunnel collapse by their activities.

In two instances badgers have been noted in the vicinity of cattle facilities. In March 1990 badger tracks, with unusually long clawprints, were noted near a cattle trough, near sett H. In November 1990 a radio-tagged female (no. 30) visited a milking parlour on the first night of radio-tracking and on two subsequent nights. The animal may well have visited the parlour on other occasions. These recorded visits occurred while dogs and people, and on one occasion a working tractor, were nearby. There was also a bull confined within an area of the parlour at the time. No observations were made while the badger was within the parlour. To date, tests using both faeces and blood have failed to detect tuberculosis from no. 30. No visits were recorded on five further nights of radio-tracking in November and December, possibly owing to the fact that a drain was dug across the gate giving access to the parlour. When this was refilled the cattle were confined in an area outside the parlour, which would have made any such visit difficult for a badger.

An adult male (no. 7) and female badger (no. 5), both infected, from the same social group (A) have been radio-tracked. Both foraged on both pasture and in damp woods/stream margins in different areas. Neither came close to cattle. No. 7 drank several times from a stream which was also used by cattle from its other side in the dry summer of 1989. As no. 5 was killed in August 1989 and no. 7's radio-tag malfunctioned in 1990, no further observations of the behaviour of infected badgers during summer are available. Badger no. 7 was not known to have TB at that time.

Badgers, when detected, are usually in the main setts of their own social group during the day, and normally were radio-tracked up to approximately 1km from the main sett at night. However, there are persistent important exceptions to this pattern. Firstly, in the dry summer of 1989 radio-tagged adult badgers (n=4) were frequently absent from their main setts (Table 2). In a total of 43 daytime checks on presence of adults in two main setts, males were found only on eight occasions and females on fourteen. Were these badgers actually absent or just undetected by radio-location? It was not always possible to locate the badgers. However, on several occasions it was possible, and it is only for these badgers that the absences are taken to be real (see methods). When absent from main sett A, the female no. 5, who was seen with a cub in that year, was located at a subsidiary sett (*c.* 500m from the main sett) on six occasions. The radio-tracked male no. 7 from sett A was located at the edge of his territory, about 1km from his main sett, in an outlier on one occasion and in a subsidiary sett on two other occasions.

Table 2. Detected presence of adult badgers in main setts during summer 1989.

MAIN SETT	MONTHS	DAYS CHECKED	Males		Females	
			Days present	Days absent	Days present	Days absent
Sett A	May, June and July	31	6	3	9	6
Sett E	June and July	12	2	3	5	3
	Total	43	8	6	14	9

Radio-location, trapping and road traffic accidents have also detected considerable long-distance (up to 8km), and sometimes extra-territorial, movements by some badgers, involving five adult females and two adult males. These cases were recorded in all seasons of the year. The long-distance movements have been confined to badgers in the smaller social groups. For example, none has been recorded for the badgers from sett A (Sleeman 1992).

DISCUSSION

These preliminary results suggest that, in this study area at least, badgers which encountered cattle, in normal circumstances, took evasive action. This agrees with the findings of Benham (1993) and Benham and Broom (1989). However, other workers have noted badgers being tolerant of/tolerated by cattle. For example, Kruuk (1989, 17) refers to a badger 'ignoring cows...often at less than a few feet distant' in Oxfordshire. However, this happened only on three occasions and only when the cattle were lying down, ruminating (Kruuk, pers. comm.). Similarly, other workers in England have noted badgers coming close to cattle on pasture (Brown *et al.* 1993). It is possible that when both badgers and cattle are present in high numbers, and for prolonged periods, a degree of habituation may occur; such habituation would be worthy of further study.

There could be a risk of cross-infection when badgers use cattle facilities such as water troughs and milking parlours. The elongated claws of the individual whose tracks were found near the drinking trough in this study suggest that this animal might have had tuberculosis. A highly significant relationship between elongate claws and the presence of tuberculosis has been detected in male Irish badgers (Whelton 1991). It is possible that badgers with tuberculosis have an increased need for water. However, such events would have to occur frequently for infection to occur, and it is considered unlikely to result in a major outbreak of tuberculosis in cattle. Contamination of pasture by infected badgers must occur, but this is still little understood (see Brown *et al.* 1993).

Dead badgers on pasture have been recorded in this study, and the investigation of such animals by curious cattle could result in a major outbreak of tuberculosis in cattle if the badger was infectious (see also Dolan 1993; Fagan 1993; Flanagan 1993). Dead (or dying) badgers in situations where they are accessible to cattle are recorded in several Irish reports on tuberculosis in cattle (e.g. O'Connor and O'Malley 1989). Today most herds of Irish cattle are housed together in winter. Investigations of dead badgers by cattle are likely to occur in spring, a time of year when it is known that tuberculosis outbreaks peak in Irish cattle (Downey 1990). It is proposed to investigate the reaction of cattle to dead badgers by presenting a suitable model badger to cattle in a variety of situations and monitoring their reactions. Because at present no dead badger can be guaranteed to be free of tuberculosis, a specially designed stuffed badger has been made.

There is evidence that cubs harbour a considerable proportion of a badger population's tuberculosis (MAFF 1990). The cubs that died at the setts in this study might well have been infectious but they were not found quickly enough to check this. We need to establish how frequently infectious badger carcasses are accessible to cattle. It is therefore planned to trap cubs twice-yearly, once in summer as they emerge and again in autumn. At the autumn trap-up they would be radio-collared, with mortality tags. These tags would give off a special signal when the animal stopped moving (i.e. indicating death), and could be fitted to standard badger radio-tags. Using these, it would be possible to locate such cubs quickly and monitor their carcasses for tuberculosis. If necessary, the study area will be expanded in order to involve sufficiently large numbers of cubs.

Despite use of trapping and snaring, it is still not certain that all the badgers in an area have been captured. Untagged animals appear from time to time, and the extent to which these are migrants or undetected resident badgers is not known. Nevertheless the density of the badger population under study is undoubtedly low in comparison to some areas of England, for example the much-studied high-density population at Woodchester Park in the Cotswolds (Neal and Cheeseman 1991). Probably as a result of this the study population has less well-defined social groups, such as have been inferred from other low-density populations, for example in the New Forest and Essex in England (Packham 1983), but in Offaly a medium-density population is clearly territorial (O'Corry-Crowe *et al.* 1993). Cheeseman and Mallinson (1979) found that trapping success in a high-density population was highest in summer, which they suggest was due to food shortage.

The density may also affect the summer absences from the main setts, which in turn may have affected trapping success in that season. This implies a different response to food shortage between the Cotswold population and this County Cork population. The success reported with trapping in spring here is considered to be due to pre-baiting when feeding with peanuts and plastic pellets. The preponderance of females in the population might be due to a higher rate of mortality in males, such as has been found elsewhere (MAFF 1990; Neal and Cheeseman 1991). Persistent persecution is also an important feature of the population under study, and is likely to be at least in part responsible for the density. These practices are probably encouraged by some landowners who are frightened of the possibility of TB-infected badgers; however, most of the landowners involved are well disposed towards badgers and this study. It is considered likely that the fact that this study is taking place, and that it involves Wildlife Service staff visiting the area to assist in the study, has prevented some persecution.

In Sussex, badgers use their main setts most frequently in winter (Roper and Christian 1992). This may also be the case in Cork but is not known for certain. In many cases radio-tagged badgers could not be located during systematic searches; the poor trapping success at main setts in summer indicates that the badgers were probably elsewhere.

Data on the occurrence of adults in main setts in summer (Table 1) have been useful in determining the optimum time of year to vaccinate adults at main setts (McCarthy 1993). However, a number of problems with methodology remain. The most critical problem is the absence of a satisfactory blood test for tuberculosis (Nolan and Goodger 1993). Detecting tuberculosis in live badgers is otherwise very difficult. There is also a need for the development of improved radio-tags, particularly tags with greater range. Owing to the long-distance movements of badgers, some of which must be out of the study area, the radio-location of such badgers with aircraft may be necessary (Sleeman 1992).

Despite the presence of an infected badger population in the area, only one herd has proved positive on testing for tuberculosis, indicating that populations of infected badgers and healthy cattle can co-exist (Cheeseman *et al.* 1988). This must commonly be the case, given the distribution of tuberculosis in Irish badgers and the fact that most cattle do not have the disease (Dolan 1993).

ACKNOWLEDGEMENTS

We gratefully acknowledge the funding support of the study from ERAD and the Lisavaird, Bandon, Barryroe and Dairygold (Ballyclough) Co-ops. We also wish to thank the many people who have assisted this study, including Tim Anderson who did the first fieldwork, the many people in diagnostic laboratories, our colleagues in England and Northern Ireland who have been extremely helpful, and our three bursary students, Mary Lenihan, Torres Sweeney and Helen Whelton.

REFERENCES

BARRY, J.M. 1990 Bovine TB and the badger. *Journal of Biomedical Science* **1**, 127-30.

BENHAM, P.F.J. 1993 The interactive behaviour of cattle and badgers with reference to transmission of bovine tuberculosis. In T.J. Hayden (ed.), *The badger,* 189-95. Dublin. Royal Irish Academy.

BENHAM, P.F.J. and BROOM, D.M. 1989 Interactions between cattle and badgers at pasture with reference to bovine tuberculosis transmission. *British Veterinary Journal* **145**, 226-41.

BROWN, J.A., HARRIS, S. and CHEESEMAN, C.L. 1993 The development of field techniques for studying potential modes of transmission of bovine tuberculosis from badgers to cattle. In T.J. Hayden (ed.), *The badger,* 139-53. Dublin. Royal Irish Academy.

CHEESEMAN, C.L. and MALLINSON, P.J. 1979 Radio tracking in the study of bovine tuberculosis in badgers. In C. I. Amlaner and D. W. Macdonald (eds), *A handbook on biotelemetry and radio tracking* 646-56. Oxford. Pergamon Press.

CHEESEMAN, C.L., CRESSWELL, W.J., HARRIS, S. and MALLINSON, P.J. 1988 Comparison of dispersal and other movements in two badger (*Meles meles*) populations. *Mammal Review* **18**, 51-9.

CHEESEMAN, C.L., WILESMITH, J.W. and STUART, F.A. 1989 Tuberculosis: the disease and its epidemiology in the badger, a review. *Epidemiology and Infection* **103**, 113-25.

COLLINS, C. H. and GRANGE, J. M. 1987 Zootic implications of *Mycobacterium bovis* infection. *Irish Veterinary Journal* **41**, 363-6.

DOLAN, L. 1993 Badgers and bovine tuberculosis in Ireland: a review. In T.J. Hayden (ed.), *The badger,* 108-16. Dublin. Royal Irish Academy.

DOWNEY, L. 1990 *Ireland's TB problem: what can and must be achieved.* Dublin. ERAD.

FAGAN, J. 1993 Tuberculosis in badgers in Ireland: pathology. In T.J. Hayden (ed.), *The badger,* 117-22. Dublin. Royal Irish Academy.

FLANAGAN, P. 1993 Badgers and bovine tuberculosis: epidemiology and case studies. In T.J. Hayden (ed.), *The badger,* 131-8. Dublin. Royal Irish Academy.

HARRIS, S. 1982 Activity patterns and habitat utilisation of badgers (*Meles meles*) in suburban Bristol: a radio tracking study. *Symposium of the Zoological Society of London* **49**, 301-23.

KENWARD, R. 1987 *Wildlife radio tagging, equipment, field technique and data analysis.* London. Academic Press.

KRUUK, H. 1978 Spatial organisation and territorial behaviour of the European badger *Meles meles. Journal of Zoology, London* **184,** 1-19.

KRUUK, H. 1989 *The social badger.* Oxford University Press.

MCCARTHY, J. 1993 The badger vaccination trial in west Cork: progress report. In T.J. Hayden (ed.), *The badger,* 181-8. Dublin. Royal Irish Academy.

MAFF 1990 *Bovine tuberculosis in badgers.* Fourteenth report. London. HMSO.

MONAGHAN, M.L., DOHERTY, M.L., COLLINS, J.D., KAZDA, J.F. and QUINN, P. J. (in press) The tuberculosis test. *Veterinary Microbiology.*

MORRIS, R.S. and PFEIFFER, D.U. 1991 Bovine tuberculosis complicated by feral reservoir: is there an answer? *Symposium on tuberculosis.* Massey University, Palmerstown North, New Zealand.

NEAL, E.G. and CHEESEMAN, C.L. 1991 Badger. In G.B. Corbet and S. Harris (eds) *The handbook of British mammals* (3rd edn). Oxford. Blackwell.

NOLAN, A. and GOODGER, J. 1993 Diagnosis of tuberculosis in live badgers. In T.J. Hayden (ed.), *The badger,* 128-30. Dublin. Royal Irish Academy.

O'CONNOR, R. 1986 *A study of the bovine tuberculosis eradication scheme.* Dublin. Economic and Social Research Institute.

O'CONNOR, R. and O'MALLEY, E. 1989 *Badgers and bovine tuberculosis in Ireland.* Dublin. ERAD/ERSI.

O'CORRY-CROWE, G., EVES, J. and HAYDEN, T.J. 1993 Sett distribution, territory size and population density of badgers (*Meles meles* L.) in east Offaly. In T.J. Hayden (ed.), *The badger,* 35-56. Dublin. Royal Irish Academy.

PACKHAM, C. G. 1983 The influence of food supply on the ecology of the badger. Unpublished B.Sc. thesis, Southampton University.

PARISH, T. and KRUUK, H. 1983 The uses of radio tracking combined with other techniques in studies of badger ecology in Scotland. *Symposium of the Zoological Society of London* **49**, 291-9.

ROPER, T.J. and CHRISTIAN, S.F. 1992 Sett use of badgers (*Meles meles*). In I.G. Priede and S.H. Swift (eds), *Wildlife telemetry. Proceedings of the 4th European conference on wildlife telemetry,* 661-9. London. Ellis Harwood.

SLEEMAN, D.P. 1992 Long distance movements in an Irish badger population. In I.G. Priede and S.H. Swift (eds), *Wildlife telemetry. Proceedings of the 4th European conference on wildlife,* 670-6. London. Ellis Harwood.

THORNTON, P.S. 1988 Density and distribution of badgers in south-west England: a predictive model. *Mammal Review* **18**, 11-23.

WARREN, A.R. 1892 The badger in Ireland. *The Field* **80**, 726.

WHELTON, H. 1991 A study into methods of age determination in the male European badger (*Meles meles*). Unpublished B.Sc. thesis, University College, Cork.

In: Hayden, T.J. (ed.) 1993 *The badger*, 166-173. Dublin. Royal Irish Academy.

THE EAST OFFALY BADGER RESEARCH PROJECT: AN INTERIM REPORT

J. Eves

District Veterinary Office, Department of Agriculture and Food, Tullamore, Co. Offaly.

ABSTRACT

In 1989, 1990 and the first six months of 1991, 1,265 badgers were removed under licence from a $600km^2$ area of County Offaly. On gross post-mortem examination 13% of these had visible lesions of tuberculosis. The APT (reactor animals per 1000 animal tests) at the end of June 1991 was 52% less than the 1988 figure in the project area, compared with an increase of 33% in the control area where badgers have not been removed.

It is too soon, however, to draw any firm conclusions from the results to date but it should be possible to do so by the end of 1993. A more detailed statistical analysis of the historical and current data for the project and control areas will be undertaken following the completion of this third year of snaring.

INTRODUCTION

The East Offaly Badger Research Programme was undertaken in order to assess the extent to which the prevailing high levels of bovine TB in an area of County Offaly might be influenced by the controlled removal of badgers. The average number of cattle reactors disclosed each year in this area had more than doubled between the late 1970s and early 1980s and continued at this higher level with minor fluctuations during the decade. In the mid-1980s, badgers were removed from three specified local areas of County Offaly and in each case the proportion of reactor herds and the incidence of reactor animals declined within two years. Similar observations were made in other areas outside Offaly (O'Connor and O'Malley 1989). These and individual case studies where there was strong circumstantial evidence to indicate a possible involvement of the badger (see for example Flanagan 1993) provided the background to an examination of the effect of badger removal over a wider area.

The project commenced in January 1989 and is to run for a three- to five-year period. The results in the project area will be compared with a designated control area from which badgers are not removed. Apart from the removal of badgers no other special measures are being undertaken in the project area. The frequency of testing, definition of blackspots and interpretation of tests are the same as in the control area.

MATERIALS AND METHODS

The study area

The badger removal area is situated in the east of County Offaly and includes parts of Counties Laois, Kildare and Meath (Fig. 1). It has an area of 600km^2. It is comprised of a central project area surrounded by a buffer zone 1.5km wide. The buffer zone is to allow, to some extent, for the movement of badgers into the project area to be taken into account. The control area includes parts of Counties Meath, Kildare, Laois, Offaly and Westmeath. The buffer zone is surrounded by rivers which divide it from the control area. The rivers Barrow, on the southern border, and Boyne, on the eastern border, form reasonably good barriers against the routine entry of badgers into the project area, but the other rivers, i.e. the Figile, Silver, Clodiagh and Yellow rivers, are quite small and in general form only poor barriers (Fig. 2). Farming in the project area is a mixture of beef, dairying, sheep and tillage.

The total number of herds and animals in each of the areas is given in Table 1. The average herd size in each of the areas is approximately 50 cattle.

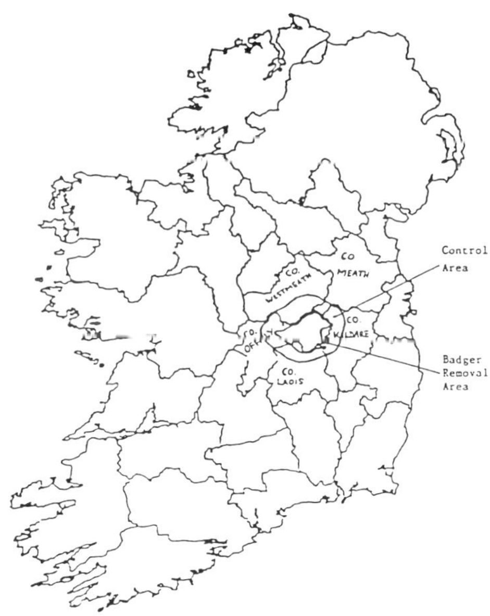

Fig. 1. East Offaly Badger Programme.

Badger removal

Between 1984 and 1988, 357 badgers had been removed from east Offaly and 56 (16%) of these were TB-positive on gross post-mortem examination. These badgers were removed under licence following investigations into cattle herd breakdowns. So, some information was available on sett location and the extent of TB infection in badgers in parts of the project area. In autumn 1988 all herd-owners in the proposed badger removal area were visited by Department staff and asked if they had badger setts on their lands. In addition, some individual areas, such as woods and the edges of bogs, were searched by Department staff.

Relying on farmers for information on the location of setts means that for a variety of reasons some setts will be missed, but lack of suitably trained personnel made it impossible to search the whole area, hedgerow by hedgerow.

Some additional setts were found in 1989 and 1990. Most of these new setts were found as a result of TB breakdowns on previously clear farms. These farms were then searched more intensively.

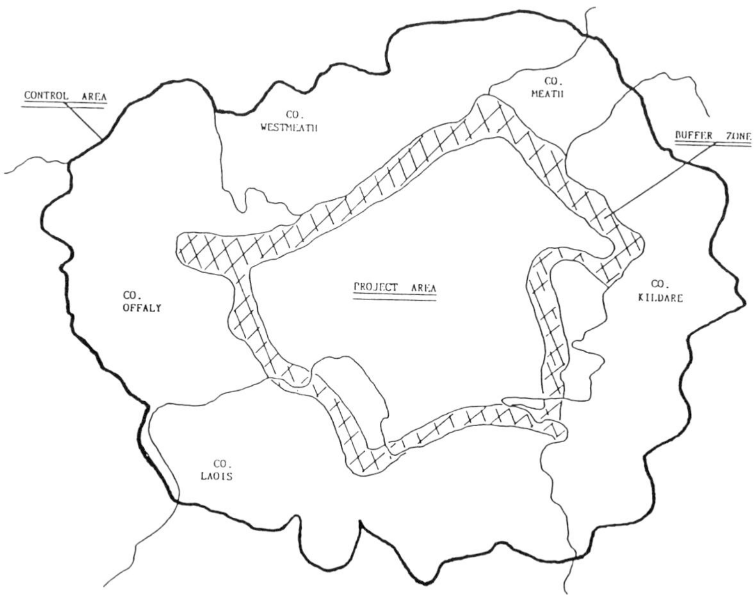

Fig. 2. Outline of the project area, buffer zone and control area of the East Offaly Badger Research Programme.

Table 1. The number of herds and the number of cattle in each of the areas associated with the East Offaly Badger Programme.

	No. of herds	No. of cattle
Project area	1027	51,490
Buffer area	330	16,954
Control area	2837	135,957
Rest of Co. Offaly	2002	107,595

Snaring technique

The snares used are in accordance with the Wildlife Act 1976 (Approved Traps and Snares Regulations 1977). They are self-locking, made from multi-strand steel wire, and are fitted with a stopper and a swivel. Snares are placed on badger pathways or 'passes' close to the sett or near the actual sett entrances. Between five and thirty-five snares may be used for each sett. They are left in position for a total of eight nights and are visited daily. Most of the badgers which are captured are caught on the first three nights. Snares are set with a large loop (200mm x 250mm) and badgers are always caught around the thorax or lower abdomen. They do not sustain any visible injury by being caught in the snare. When captured they are shot in the head with a .22 bullet.

Snaring commenced in January 1989 and continued until May. It recommenced in September and continued until December 1989. Five field staff snared in spring and four in autumn. Badger removal started on the eastern side of the area and moved westwards. All field staff were concentrated in one area at a given time and each local area was systematically surveyed prior to snaring. A similar programme was undertaken in 1990 with a staff of three. Snaring began again in 1991 with two field staff.

In the field, badger carcasses are tagged and information on the place of capture, weight, and sex is recorded. They are transported to the laboratory either on the day of capture or on the following day.

In the laboratory badgers which have TB-like lesions have samples taken for histology and culture. Tuberculosis is confirmed if one or both of these is positive. Badgers which appear negative on gross examination have no samples taken.

RESULTS

Badger setts

To date a total of 635 setts have been found. Most of these have three entrances or more and are probably main setts or subsidiary setts.

Table 2 sets out the location of these setts. Most of the setts found on farmland are in hedgerows with occasional setts located in the middle of a pasture or tillage field. Deciduous woods are very scarce in the area but where present they are usually chosen by badgers. Many of the woods listed under this heading are in fact small mixed woods. The setts located on the bog edges were generally in scrubland and situated on sloping ground. There is a large number of coniferous woods in the study area but they only contain about 5% of the setts. When setts are present in these woods they are often located very close to the perimeter. The remainder of the setts were found in such diverse places as railway embankments, river and canal banks, under roadways, graveyards, orchards, gardens, a football field and a golf course. The proportion of setts found on farmland as opposed to woodland is quite high but is probably explained by the lack of suitable woodland in east Offaly.

Badgers caught

A total of 837 badgers were caught in 1989, 300 in 1990, and 128 were caught from February to June in 1991. In 1989 the whole area was snared once; in 1990, twice; and based on progress to date, three snarings should be completed in 1991.

In spring 1989 about 70% of the setts showed signs of occupation by badgers and in spring 1991 that figure was about 20%. During the two and a half year period there has been a noticeable decline in the amount of badger activity in the area. There has also been a decline in the number of road casualties and badger sightings. However, the indications are that the complete removal of all badgers from the area, by snaring, may not be achieved.

Table 2. The number of badger setts identified and their association with landscape features.

Landscape features	Number of setts	% of all setts identified
Farmland	402	63
Woods	92	14
Bog edges	82	13
Coniferous woods	31	5
Other	28	5
Totals	635	100

Tuberculosis in badgers

Table 3 shows the numbers of badgers snared in the different areas and the numbers which were TB-positive. The prevalence of gross lesions of tuberculosis in the badgers that were captured decreased from 14% in 1989 to 9% in 1990 and has increased to 11% for the first half of 1991. In 1990 and, so far, in 1991, 34% and 31% respectively of all badgers caught in the east Offaly area were caught in the buffer zone as against 16% caught in the buffer zone in 1989. This would seem to indicate a movement of badgers into the area.

Gallagher et al. (1976) reported that 24(66%) of 36 infected badgers had gross lesions of tuberculosis Fagan (1993) has reported a similar finding in Ireland. One could reasonably speculate that the 13% gross lesion figure represents just two thirds of the real infection rate in the badgers captured in east Offaly.

Tuberculosis in cattle

The details of the animal tests, reactors disclosed and reactor animals per thousand tests (APT) are recorded in Tables 4 and 5. They show a steady decline in the APT figures for the central project area of -23% at the end of 1990 and currently of -52% (half year) compared with an increase of +10% at the end of 1990 and currently of +33% (half year) in the surrounding control area. The rest of Offaly, excluding the control area, showed a marked increase in 1989 (+53%), virtually no change in 1990 and has declined by 20% in the first half of 1991, during which badger control operations have been undertaken in areas which had high herd breakdown rates.

Table 3. The total number of badgers snared in the East Offaly Badger Programme by project area, buffer zone, number tuberculous and year.

Year		Project area	Buffer zone	Total
1989	Total snared	700	137	837
	Total positive	94(13)	25(18)	119(14)
1990	Total snared	197	103	300
	Total positive	24(12)	3(3)	27(9)
1991 (Jan-June)	Total snared	88	40	128
	Total positive	12(14)	2(5)	14(11)
	Total	985	280	1265
		130(13)	30(11)	160(13)

( ) Per cent of badgers which were TB-positive.

Table 4. The number of animal tuberculin tests, number of tuberculin reactors, the APT and change in the project, control and rest of Offaly areas of the East Offaly Badger Programme in 1988, 1989, 1990 and 1991 (half year).

Area	Year	Tests	Reactors	APT	%Change	Test per animal
Project						
	1988	86190	321	3.72		1.67
	1989	102845	343	3.34		2.00
	1990	97976	281	2.97	- 23	1.90
	1991 (to 30 June)	31247	56	1.79	- 52	0.61
Control						
	1988	249544	720	2.89		1.88
	1989	268675	829	3.09		1.98
	1990	247759	789	3.18	+ 10	1.82
	1991 (to 30 June)	81325	350	4.30	+ 33	0.60
Rest of Offaly						
	1988	197707	619	3.13		1.84
	1989	236741	1123	4.74		2.20
	1990	222293	1045	4.70	+ 33	2.07
	1991 (to 30 June)	77295	285	3.70	+ 15	0.72

The figures for the first half of 1991 show a significant reduction in the APT for the project area; however, the indications are that the APT may rise for the second half of the year. This has been the pattern of the disease in the area for a number of years whereby the autumn peak of reactors is higher than the spring peak.

Table 5. The total number of reactors and positive herds recorded during the years 1988, 1989 and 1990 in the East Offaly Badger Programme.

Year	Positive herds	Total no. of reactors	Mean no. of reactors per herd
1988	110	321	2.97
1989	123	343	2.79
1990	105	281	2.69

DISCUSSION

The prevalence of bovine tuberculosis in County Offaly has remained at a high level throughout the 1980s. In areas of the county over 40% of herds had breakdowns during a 12-month period (O'Connor and O'Malley 1989). There was considerable circumstantial evidence of the involvement of the associated tuberculous badger population in these areas. When badger removal operations were undertaken, between 10% and 40% of the badgers were shown to have gross lesions of tuberculosis. The prevalence of bovine tuberculosis in the associated cattle dropped sharply, following a lag period, once the infected badgers were removed.

The present project was undertaken to investigate and quantify the impact of systematic trapping of badgers, from a large area, on the prevalence of bovine tuberculosis in the associated cattle population. The results would be compared with the disease prevalence in the surrounding control area where badgers were not being trapped. A buffer zone was incorporated between these two areas to allow for immigration of badgers into the area where snaring had been carried out.

The process of trapping badgers by snaring was expected, initially, to be about 60%-70% successful. Setts were revisited on a systematic basis and where activity was observed, snaring was repeated. O'Corry-Crowe *et al.* (1993) have estimated that 51% of adult badgers were trapped or died during the first snaring effort in a defined study area within the main project area. Following repeat snaring they estimated that 80%-90% of the adult population in the area had been trapped or had died by the autumn of 1990. Currently, it would appear that the number of active setts has been reduced by approximately 80% from the 1989 level, but there is evidence of constant immigration of badgers to augment the residual untrapped animals.

The prevalence of tuberculosis in the cattle in the project area, from 1988 to the end of June 1991, has decreased by 52%, based on the APT figures, while the prevalence in the control area has increased by 33%. While the figures show a strong and consistent trend in the project area in contrast to the increase in the control area, it is too soon to draw any firm conclusions from the results to date.

In order to analyse fully the effect of badger removal on the prevalence of tuberculosis in the cattle population it will be necessary to establish the disease pattern prior to badger removal and compare that with the subsequent disease pattern. Statistical methods, such as the Proportional Hazards Model, will be used to assess, over time, whether herds in the project area have a lower risk of contracting tuberculosis than herds in the control or buffer areas. The data will also be subjected to analysis using geographical information systems. The map coordinates for each badger sett and farm are being recorded so that the relationship between breakdown herds and infected badger groupings can be assessed. Such analytical procedures provide the best method of evaluating and quantifying the contribution of the tuberculous badger population to the prevalence of tuberculosis in the associated cattle in the situation where the procedure of snaring is less than 100% successful.

ACKNOWLEDGEMENTS

I wish to thank Mr D.P. Bray, S.V.I., District Veterinary Office, Tullamore, for his constant support during the project. I also wish to thank Mr Michael Hennessy, Agricultural Officer, who has supervised much of the fieldwork. My thanks are also due to members of the Tuberculosis Investigation Unit for tuberculin test data. I am also grateful to the staff of the Veterinary Research Centre, Abbotstown, and the Regional Veterinary Laboratories in Athlone and Limerick who carried out the post-mortem examinations on the badger carcasses.

REFERENCES

FAGAN, J. 1993 Tuberculosis in badgers in Ireland: pathology. In T.J. Hayden (ed.), *The badger,* 117-22. Dublin. Royal Irish Academy.

GALLAGHER, J., MUIRHEAD, R.H. and BURN, K.J. 1976 Tuberculosis in wild badgers (*Meles meles*) in Gloucestershire: pathology. *Veterinary Record* **98**, 9-14.

FLANAGAN, P. 1993 Badgers and bovine tuberculosis: epidemiology and case studies. In T.J. Hayden (ed.), *The badger,* 131-8. Dublin. Royal Irish Academy.

O'CONNOR, R. and O'MALLEY, E. 1989 *Badgers and bovine tuberculosis in Ireland.* Dublin. Eradication of Animal Diseases Board (ERAD).

O'CORRY-CROWE, G., EVES, J. and HAYDEN, T.J. 1993 Sett distribution, territory size and population density of badgers (*Meles meles* L.) in east Offaly. In T.J. Hayden (ed.), *The badger.* 35-56. Dublin. Royal Irish Academy.

In: Hayden, T.J. (ed.) 1993 *The badger*, 174-180. Dublin. Royal Irish Academy.

COULD VACCINATION BE THE ANSWER TO THE BADGER TUBERCULOSIS PROBLEM?

J. Stanford,† K. Stainsby* and K.H. Mahmood **

Department of Medical Microbiology, University College and Middlesex School of Medicine School of Pathology, Riding House Street, London WIP 7LD

*** Present address: Department of Microbiology, University of Surrey, Guildford.**
**** Present address: Department of Animal Production, University of Basrah, Iraq.**

†To whom correspondence should be addressed.

ABSTRACT

Badger tuberculosis requires a long-term solution. Options for reducing the spread of the disease from badgers to cattle are reviewed, and the possible utility of an oral vaccine for badgers is considered. Mammalian cell-mediated immunity to mycobacterial disease as found in the badger is summarised. The ability of the badger to survive long into the terminal anergic phase of tuberculosis explains why the animal constitutes such an important source of infection. The threefold potential of a vaccine and the special requirements in relation to the badger are discussed. The background to the selection of killed *Mycobacterium vaccae* as a potential vaccine is briefly explained, and the simple procedure for the production of Badge-vac is outlined.

INTRODUCTION

Tuberculosis in badgers

Badger tuberculosis is here to stay and, as long as husbandry practices allow cattle to mix with or share their pastures with badgers, transmission to cattle will continue to be a problem. Means of solving the problems are few, but range from encouraging agricultural practices other than cattle farming in areas where badgers are common to the eradication of the badger. The reasonable choice lies between these two extremes.

Options under consideration include relatively minor changes in husbandry to reduce contact between cattle and badgers, and the reduction of badger densities in selected areas to try to limit the spread of tuberculosis between badgers. Other potential options under special circumstances include development of a test to detect tuberculous animals which could be selectively culled (Nolan and Goodger 1993). However, a more widely applicable approach would be the development of a vaccine that could be safely applied to badgers cheaply and with minimal disturbance of the environment. The vaccination option is less complicated than it sounds; this paper and that by McCarthy (1993) report on the development of a potential vaccine, and on its field application.

How the problem has been dealt with in other species

The method generally used to combat tuberculosis in humans and cattle is the removal of infected individuals. In man, modern treatment effectively makes an infectious person non-infectious in about a week. In cattle, regular tuberculin testing allows removal of reactors which include most, but perhaps not all, infectious animals. Application of either of these approaches is not feasible in badgers except in limited circumstances. Vaccination with BCG has been useful in controlling tuberculosis in man, and between 70% and 80% protection from disease is achieved in the UK and Scandinavia (Medical Research Council 1956; Tverdal and Funnemark 1988). If even 50% protection could be achieved in badgers, spread of the disease amongst them and to cattle would be markedly reduced.

IMMUNOLOGICAL CONSIDERATIONS

Immunity and mycobacteria

The mammalian immune system combats infection with mycobacteria such as the tubercule bacillus with a complicated combination of cellular activities (Mackaness 1968). In brief these include expansion of clones of thymic (T) lymphocytes that recognise particular mycobacterial antigens presented on the surface of phagocytotic cells of the macrophage series. Cytokines released by these T cells induce the macrophage to become activated, and it is believed that activation of a particular type leads to the production of enzymes by the macrophage that directly or indirectly kill engulfed bacilli. Simultaneously with this cellular immune mechanism, expanded clones of bursal (B) lymphocytes produce high titres of antibodies to surface and secreted antigens of tubercule bacilli (Abou-Zeid *et al.* 1988), and perhaps neutralise tissue-damaging attributes of some antigens (Ratcliffe *et al.* 1988).

If at the time of bacterial challenge this protective immunity has not developed or if it is overcome, either because of a large challenge dose or because the immune status of the animal has been temporarily undermined by stress, then tuberculosis disease may develop. Development of disease is accompanied by a change in the cellular immunological mechanism to one in which the host cells containing mycobacteria, and expressing mycobacteria antigens on their surface, are killed in response to a different set of cytokines (Rook *et al.* 1987; 1989). In this state of disease, it is the host cells contaminated with bacilli that are killed, not the bacilli themselves. Thus lesions of the disease consist of abscesses of dead tissue, containing numerous live bacilli, which are surrounded by a collar of macrophages, lymphocytes, and fibrous tissue. Although disease is limited by the altered immunity, it is not totally overcome, and the animal may live for months or years with fluctuating disease, becoming infectious to others whenever the tuberculous abscesses erode into the respiratory, renal or intestinal tract.

Is the badger a special case?

At first it was believed that badgers did not have the normal mammalian pathways of immunity (Morris *et al.* 1978), because many of them were seen in the rather unusual immunosuppressed end-stage of disease when there is very little immunological response to overwhelming numbers of organisms (|Little *et al.* 1982). It is now known that the badger does make normal mammalian immune responses (Mahmood *et al.* 1987 a, b), but has the unusual ability to survive extremes of the disease that would kill most other species and to continue living long into the stage of failed immunity (Thorn and Morris 1983). It is this remarkable capacity of badger survival that makes it such a danger to other species susceptible to infection with tubercule bacilli.

This survival under conditions that would kill most other species may be due to some resistance of the badger cells to the toxic effects of cord factor from the tubercule bacilli. In man and mouse, cells containing more than five or six mammalian tubercule bacilli die, whereas the same cells would remain viable if they contained many times more *Mycobacterium avium* (avian tubercule bacilli). Thus *M. avium* infections in humans, and in mice, often progress to a situation similar to that of the badger with advanced tuberculosis.

Except for perhaps possessing unusually resistant cells, the badger responds to tuberculosis much as other mammals. Like other mammals, too, healthy badgers produce antibodies against the mycobacterial species they meet in their environment (Stainsby *et al.* 1989). In view of this, it seems reasonable to draw parallels for the badger from what is known of tuberculosis in much more easily studied animals such as man.

The psychosis of advanced tuberculosis

It was well known in Victorian times that persons suffering from advanced consumption did not always behave in a rational way, and sometimes showed more than ordinary brilliance in their last days. Fortunately we rarely see this terminal psychosis in developed countries.

That the psychosis of advanced tuberculosis affects badgers is seen in the aberrant behaviour pattern often assumed by highly infectious individuals (Cheeseman and Mallinson 1981). Such animals wander outside their normal territories, sometimes for miles, potentially infecting the badger communities they encounter en route. They also tend to be seen in daylight, and to sleep in sheds and byres, contaminating the surroundings of cattle.

A VACCINE FOR BADGERS

What would a vaccine do?

A vaccine against tuberculosis can potentially exert three separate effects. (1) A true vaccine effect induces protective immunity, the efficacy of which may be influenced by the animal's state of immunity at the time of vaccination and a number of environmental factors. These include contact with free-living mycobacteria, and a variety of stresses. (2) The vaccine may have a preventative immunotherapeutic effect leading to the conversion of the tissue-damaging type of response to protective immunity. This may result in immune destruction of any tubercule bacilli surviving in the tissues so that subsequent stresses will not lead to the development of active disease. (3) The vaccine may also have an active immunotherapeutic effect on animals in the early stages of tuberculosis, modifying its course, and perhaps aborting its extension.

The constraints to be overcome in producing a badger vaccine

For obvious reasons a potential vaccine for badgers must not only fulfil the immunological requirements but be cheap, easy to use, and suitable for oral administration. It must not damage the environment, or have unwanted effects on other species. Thus we are limited to simple formulations of a killed vaccine strain presented in a way that will be readily taken up by badgers. The potential vaccine must be safe if taken up in many doses by badgers or cattle whether on a single occasion or on a number of separate occasions.

Most of the limiting factors have been, or are being, overcome, and a suitable preparation known as Badge-vac has been produced and is undergoing field evaluation.

Selection of a potential vaccine strain

The vaccine used to prevent tuberculosis in man, the bacillus of Calmette and Guerin, (BCG), consists of a suspension of living bacteria which is effective if taken by mouth although it is normally given by intradermal injection nowadays. History tells us that BCG is an attenuated form of the bovine strain of the tubercule bacillus (Grange *et al.* 1983). BCG vaccination can occasionally lead to disease in immunosuppressed individuals (Lotte *et al.* 1984), and directly produces a high level of tuberculin positivity. Since the latter is the basis of the tuberculosis eradication programme for cattle, a vaccine could not be left around the countryside that could inadvertently be eaten by cattle and lead to their tuberculin conversion. Equally, it is impossible to be sure that if BCG, although attenuated, were to be released into the environment it would not revert to a more virulent form. Unfortunately, when killed, BCG is much less effective and some alternative is needed for a badger vaccine.

Since 1978, killed suspensions of a fast-growing environmental species *Mycobacterium vaccae* have been under investigation in man for their vaccinating capacity against leprosy. Such suspensions have been used alone or as additives to BCG (Ganapati *et al.* 1989; Ghazi-Saidi *et al.* 1989). Subsequent investigation has shown that killed *M. vaccae* can be used with success as an immunotherapeutic adjunct to chemotherapy in the treatment of tuberculosis in man (Bahr *et al.* 1990; Stanford, Bahr *et al.* 1990).

Could M. vaccae *be the basis of a vaccine for badgers?*

The indirect evidence from man and from various models of disease in experimental animals is now overwhelming. There can be little doubt that killed *M. vaccae* is a very potent agent for the prevention and treatment of tuberculosis and leprosy, and data are accumulating on its effects in rheumatoid arthritis and psoriasis. Direct evidence for its efficacy in the badger is much less, and its collection is so full of potential pitfalls that there is probably little point in trying to engender it from experiments in captive badgers. Thus the present study in Ireland aims to vaccinate badgers and observe the secondary effect that this may have in reducing tuberculosis in the cattle herds in contact with the badgers.

One great advantage of *M. vaccae* is that it does not directly induce tuberculin positivity, and therefore should not confuse the diagnostic value of this test in cattle herd surveillance. Another great advantage is that not only does *M. vaccae* immunise with the protective common mycobacterial antigens, but it also has the ability to switch immune mechanisms from the tissue-damaging type of response to the protective immune response. Thus it should also be effective in both preventive immunotherapy and treatment of early disease.

Summary of the preliminary experiments

After establishing that badgers had fundamentally normal mammalian responses to mycobacteria (Mahmood *et al.* 1987 a, b), two joint studies were set up between the Central Veterinary Laboratory at Weybridge and ourselves. The first of these established that BCG vaccine did not cause progressive disease in badgers, and that it induced responses similar to those found in man. The second showed that vaccinating badgers with BCG or *M. vaccae* by a number of routes altered their response to subsequent injected challenges with virulent bovine tubercule bacilli. This pilot study showed that an oral dose of ten killed *M. vaccae* was more effective than the other vaccines tested in limiting the survival of badgers in the anergic and highly infectious terminal phase of the disease.

The pilot studies also underlined the many problems of such experiments with captive wild animals, and persuaded the authors that more extensive investigations along these lines would not be useful. This was due in part to the varied backgrounds and ages of such animals, the difficulty in avoiding stresses that would undermine the results, and the necessity for using an unrealistic challenge dose and system.

Before the vaccine could be applied in the field, it was necessary to study the effects it might have on the tuberculin test in cattle. Experiments carried out in Dublin and in Weybridge showed that multiple oral doses of the proposed vaccine did not induce tuberculin positivity. Experiments are presently in progress at Weybridge to discover whether tuberculin-positive cattle would be converted to negativity by large doses of the vaccine. When this was tried in a single tuberculous animal in Dublin, suppression of the test did not occur.

The making of Badge-vac

Having decided upon the active principle of the vaccine, the problems of dosage, presentation and administration remain. The vaccine must be presented so that it is taken preferentially by the badger rather than other animals, and so that neither too low nor too high a dose is taken.

The doses of *M. vaccae* used in man range from 10^7 to $2x10^9$, given by injection. For vaccination of persons unsuitable for BCG, the dose of 10^8 *M. vaccae* given alone has been successful (Stanford, Bahr *et al.* 1990), and the higher doses have been reserved for immunotherapy.

Assuming that at least ten times the dose would be required by mouth than is needed by injection, 10^9 was selected. Experiments with mice showed that up to 200 such doses were required before immunosuppression was induced.

For a vaccine that has to be left as bait in the environment, a dry preparation was chosen. The dose of 10^9 autoclaved bacilli is mixed with potato starch and spray-dried lactose, and packed into size number 2 gelatine capsules. These capsules are then coated in chocolate for the trial near Cork.

REFERENCES

ABOU-ZEID, C., RATCLIFFE, T.L., WIKER, H.G., HARBOE, M., BENNEDEEN, J. and ROOK, G.A.W. 1988 Characterisation of fibronectin-binding antigens released by *M. tuberculosis* and *M. bovis* BCG. *Infection and Immunity* **56**, 3046-51.

BAHR, G.M., SHAABAN, M.A., GABRIEL, M. *et al.* 1990 Improved immuno-therapy for pulmonary tuberculosis with *Mycobacterium vaccae. Tubercule* **71**, 259-66.

CHEESEMAN, C. L. and MALLINSON, P. J. 1981 Behaviour of badgers (*Meles meles*) infected with bovine tuberculosis. *Journal of Zoology* **194**, 284-9.

GANAPATI, R.P., REVANKAR, C.R., LOCKWOOD, D.N.J. *et al.* 1989 A pilot study of three potential vaccines for leprosy in Bombay. *International Journal of Leprosy* **57**, 33-7.

GHAZI-SAIDI, K., STANFORD, J.L., STANFORD, C.A., DOWLATI, Y., FARSHCHI, Y., ROOK, G.A.W. and REES, R.J.W. 1989 Vaccination and skin test studies on children living in villages with differing endemicity for leprosy and tuberculosis. *International Journal of Leprosy* **57**, 45-53.

GRANGE, J.M., GIBSON, J.A., OSBORN, T.W., COLLINS, C.H. and YATES, M.D. 1983 What is BCG? *Tubercule* **64**, 129-39.

LITTLE, T.W.A., NAYLOR, P.F. and WILESMITH, J.M. 1982 Laboratory studies of *M. bovis* infection in badgers and calves.*Veterinary Record* **111**, 550-7.

LOTTE, A., WASZ-HOCKERT, O., POISSOU, N., DUMITRESCU, N. and VERRON, M. 1984 Complications caused by BCG vaccination: retrospective studies. *Bulletin of the International Union against Tuberculosis* **55**, 55-64.

MCCARTHY, J. 1993 The badger vaccination trial in west Cork: progress report. In T.J. Hayden (ed.), *The badger,* 181-8. Dublin. Royal Irish Academy.

MACKANESS, G.B. 1968 The immunology of antituberculous immunity. *American Review of Respiratory Diseases* **97**, 337-44.

MAHMOOD, K.H., ROOK, G.A.W., STANFORD, J.L., STUART, F.A. and PRITCHARD, D.G. 1987a The immunological consequences of challenge with bovine tubercule bacilli in badgers (*Meles meles*). *Epidemiology and Infection* **98**, 155-63.

MAHMOOD, K.H., ROOK, G.A.W., STANFORD, J.L., STUART, P.A., PRITCHARD, D.G. and BREWER, J. 1987b The immune response in two populations of wild badgers naturally infected with bovine tubercule bacilli. *Tubercule* **68**, 119-25.

MEDICAL RESEARCH COUNCIL 1956 BCG and vole bacillus vaccines in the prevention of tuberculosis in adolescents. First report. *British Medical Journal* **1**, 413-27.

MORRIS, J.A., STEVENS, A.E., LITTLE, T.W.A. and STUART, F. 1978 Lymphocyte unresponsiveness to PPD tuberculin in badgers infected with *M. bovis. Research in Veterinary Science* **25**, 390-2.

NOLAN, A. and GOODGER, J. 1993 Diagnosis of tuberculosis in live badgers. In T.J. Hayden (ed.), *The badger,* 128-30. Dublin. Royal Irish Academy.

RATCLIFFE, T.L., MCGARR, J.A., ABOU-ZEID, C., ROOK, G.A.W., STANFORD, J.L., ASLANZADEH, J. and BROWN, E.J. 1988 Attachment of mycobacteria to fibronectin coated surfaces. *Infection and Immunity* **134**, 1307-13.

ROOK, G.A.W., AL-ATTIYAH, R. and FOLEY, N. 1989 The role of cytokines in the immunopathology of tuberculosis and the regulation of agalactosyl IgG. *Lyphokine Research* **8**, 323-8.

ROOK, G.A.W., TAVERNE, J., LEVETON, C. and STEELE, J. 1987 The role of gamma interferon, vitamin d3 metabolites and tumour necrosis factor in the pathogenesis of tuberculosis. *Immunology* **62**, 229-34.

STAINSBY, K.J., MAHMOOD, K.H. and STANFORD, J.L. 1989 Antibodies to mycobacteria of healthy and tuberculous badgers from two English counties. *Tubercule* **70**, 287-92.

STANFORD, J.L., BAHR, G.M., ROOK, G.A.W. *et al.* 1990 Immunotherapy with *Mycobacterium vaccae* as an adjunct to chemotherapy in the treatment of pulmonary tuberculosis. *Tubercule* **71**, 87-93.

STANFORD, J.L., STANFORD, C.A., GHAZI-SAIDI, K. *et al.* 1989 Vaccination and skin test studies on the children of leprosy patients. *International Journal of Leprosy* **57**, 38-44.

THORN, C.J. and MORRIS, J.A. 1983 The immune spectrum of *M. bovis* infections in some mammalian species:a review. *Veterinary Bulletin* **53**, 543-50.

TVERDAL, A. and FUNNEMARK, E. 1988 Protective effect of BCG vaccination in Norway 1956-73. *Tubercule* **69**, 119-23.

In: Hayden, T.J. (ed.) 1993 *The badger*, 181-188. Dublin. Royal Irish Academy.

THE BADGER VACCINATION TRIAL IN WEST CORK: PROGRESS REPORT

J. McCarthy

Department of Agriculture, District Veterinary Office, South Mall, Cork

ABSTRACT

Badger vaccination is a new concept in the control of tuberculosis (TB) in badgers. The vaccine is administered orally to badgers by including it in bait, placed on tracks, near their setts. Initial difficulties in the delivery of the vaccine to the badgers have now been overcome. Capsules containing the vaccine are now chocolate-coated, and are readily consumed. The purpose of the five-year programme in west Cork is to monitor the effects, if any, of vaccination on TB levels in badgers and associated cattle herds. It is now generally acknowledged that infected badgers may transmit TB to cattle, and account for a significant proportion of multiple reactor breakdowns. Should vaccination prove successful, it will be of immense value not only in preserving the badger population but also in keeping cattle herds free of TB. The actual method of vaccination is simple, and, over a wide area, would be carried out in a fraction of the time and cost involved in other methods of reducing TB in badgers.

INTRODUCTION

While large-scale field trials of the effectiveness of vaccination of foxes against rabies have been reported, vaccination of badgers against *M. bovis* is a new concept in the control of tuberculosis (TB) in badgers (Brochier *et al.* 1991). A programme of oral vaccination of badgers against TB, using killed *Mycobacterium vaccae*, was instituted in west Cork in June 1988. The vaccine was developed at the University College and Middlesex School of Medicine, London (Stanford *et al.* 1993). This trial is funded by ERAD, the government agency appointed by the minister of agriculture and food for the eradication of animal diseases.

Criteria used in the selection of the experimental area were that all areas selected for inclusion in the field trial have (1) a history of chronic infection in cattle herds; (2) TB confirmed in captured badgers; (3) herd breakdowns attributable to badgers; and (4) failure of tuberculin testing and routine procedures to prevent recurring breakdowns.

The objectives of the field trial are (1) to establish efficient methods of administering oral vaccine to a population of wild badgers; and (2) to determine the effects of annual vaccination, over a five-year period, of *Mycobacterium bovis*-infected badger populations on the incidence of TB in cattle.

At first, the trial was carried out in a small area during 1988 and 1989. Subsequently, the area was expanded to include an area with better natural boundaries. Thus, the trial proper commenced in spring 1990. By this time, measures for the delivery of vaccine to the badgers had been perfected.

The current programme includes three experimental areas: a study area and two control areas. Badgers in the study area will receive vaccine. It is also proposed to blood-sample some badgers, either during or after the trial, to determine what percentage of the population has been vaccinated. Cattle reactor trends will be contrasted for all three areas. At present, no firm conclusions can be drawn, as the trial has only been in progress for one year.

MATERIALS AND METHODS

The study area

The study area (i.e. vaccination area) covers approximately 200km^2, is situated in the Clonakilty area of west Cork (55km from Cork City), and is divided across the centre by a main road (Fig. 1). The coastline forms the southern boundary; the northern boundary is part mountain and bog, with a river along its entire length. The boundaries to the east and west are short over land because of deep sea inlets and contain natural barriers to wildlife movement. Thus the area is almost entirely free of any normal intrusion of badgers from outside. The altitude increases from south to north, and ranges from 30m to 150m above sea level. The valleys curve in an east to south-west direction, forming a landscape of rolling lowlands. The soil is light and sandy near the coast: a gravel vein courses through its northern portion. The area is predominantly pastureland, with a few areas of fodder crops, and some woodland (= 100ha) of mixed broad-leaved and coniferous trees. Farming is mainly dairying with a capacity to rear drystock for the store market. The herds are self-contained, replacement stock being retained while surplus stock, with few exceptions, is sold at local marts for trade outside the county.

Sett location and distribution

Approximately 160 setts, varying in size and activity, have been identified. Normally, about a third of these setts are active when examined twice-yearly for vaccination purposes. In 1990, 64 active setts were identified. This gives an active sett density of 31 per 100km^2. In Avon, Cornwall, and Gloucestershire, a sett density of 30-50 per 100km^2 was found (Cheeseman *et al.* 1981). The majority of these setts are on pastureland. A high proportion exist in association with round forts or earthworks located on grazing ground. These fortresses of 'safety' for badgers date back to 4000 B.C. and are rarely interfered with by farmers because of local superstition and folklore.

Badger density

A National Badger Survey is currently being undertaken by the Wildlife Service (Smal 1993). Meanwhile, as such information is lacking for County Cork, an estimate of badger density was derived from snaring results from a number of areas within west

Cork. A badger density of approximately 2.5 per km^2 was found (no distinction being made between juveniles and adults). In contrast, the adult population was 4.7/km^2 in Avon and Cornwall, 6.2/km^2 in Staffordshire, and 20/km^2 in the Cotswolds of Gloucestershire (Cheeseman *et al.* 1981; 1985). It has been suggested that the critical density for badger-induced disease is one infected badger per km^2, which indicates that at a rate of 2.5/km^2 *Mycobacterium bovis* could persist at moderate levels (Anderson and Trewhella 1985).

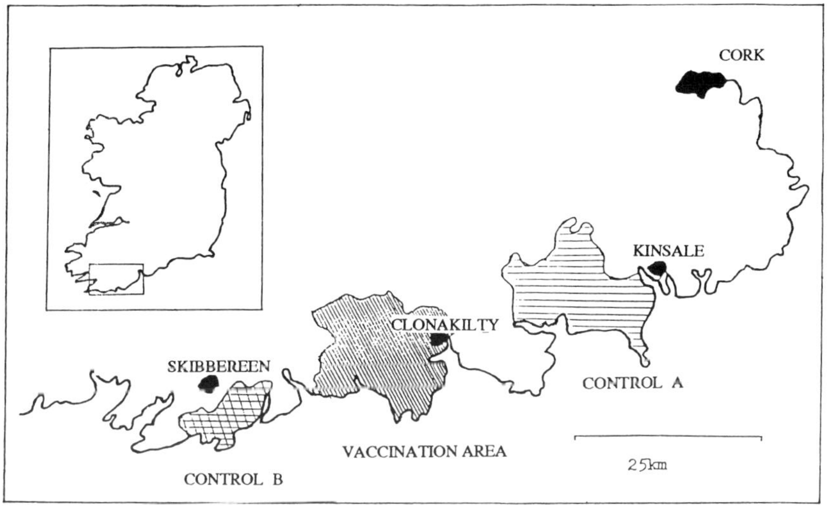

Fig. 1. Map of the three experimental areas.

Badger TB prevalence

Prior to the start of the pre-trial programme, snaring was conducted at a low level, to control badger-related breakdowns in cattle. In addition, to establish the extent of disease in the badgers, a further sixteen animals were snared. On culture, eight of these were positive for TB. Available data would suggest an overall disease level of 20%. This prevalence compares with a national level of 17% and an 18.5% level in south-west Cork based on post-mortem results compiled over a ten-year period, 1980-1989 (Dolan 1993).

Delivery of the vaccine

The vaccine is presented in gelatin capsules, containing 10,000 million *Mycobacterium vaccae* (killed by autoclaving), together with starch and lactose. The proposed name for the preparation is 'Badge-vac'. It is an oral vaccine for use in badgers.

At first, the delivery of the gelatin-coated capsules to the badgers was only partially successful. They were to a certain extent unappetising, some being totally rejected while others were partially rejected, as evidenced by the presence of tooth marks on the capsular remains. However, this problem was overcome by chocolate-coating the capsules. The badgers' sense of smell is extraordinarily well developed and they are able to discriminate between a wide variety of different scents (Neal 1986). They relished the chocolate. The capsules were presented in portions as in a typical bar of chocolate ($2cm^2$).

Bait-laying

The vaccine is delivered to the badgers in baits of a peanut and molasses mixture. Baits are placed along active tracks at 1m intervals, close to the sett entrances, on two consecutive days and on a third day during the following week. Ten to fifteen baits are placed per sett at each visit. A shallow hole ($20cm^2$ approx.) is dug using a small spade, and into it is deposited a dessertspoonful amount of bait mixture, topped up with one chocolate-coated capsule. The holes are covered by the removed sods or preferably, stones large enough to deter non-target species such as pheasant, fox, rat, and field mouse. All baited setts are routinely checked 24 hours later to determine the success or failure of the delivery to the badgers. Typical signs of a successful delivery include rooting, badger prints, faeces with undigested pieces of peanuts, and the absence of capsules.

Cattle reactor trends

Effective vaccination would be extremely useful in the control of badger TB and its secondary effect may be to reduce TB in associated cattle herds (Stanford *et al.* 1994). The following hypothesis may be said to either validate or invalidate the secondary effect, i.e. the reduction of TB in cattle by badger vaccination: 'Infected badgers transmit TB to cattle which results in multiple breakdowns in herds, with a high lesion rate'. In conjunction with the study area, two control areas (A and B) have been selected to test the hypothesis. They are coastal and have similar topographical features and farming practices to the study area (Fig. 1).

Control A is almost equal in size to the study area, has a chronic TB problem, and a similar badger density. The disease level has ranged from a high of 220 to a low of 18 reactors over the past ten years (Fig. 2). It should continue to cycle and therefore provide a comparison to the study area. The badgers of this area will not be subjected to any intentional interference. Control B is approximately one-fifth the size of the study area. It has had a long-term TB problem. Since 1986, badgers have been snared each year in this area. To date, 90 badgers have been removed, of which 21 (23%) were positive on examination post-mortem. The reactor rate in cattle is now almost nil in this area (Fig. 2). If this condition persists, it might suggest that the absence of TB in the cattle herds is due to the elimination of badgers within the area. Thus, Control B serves two purposes: (1) it validates part of the hypothesis, i.e. 'infected badgers transmit TB to cattle', and (2) as badger vaccination is intended to eliminate the badger factor either wholly or in part, it provides a falling reactor rate over five years, a trend that the study area may have to, or very nearly, attain should vaccination prove successful.

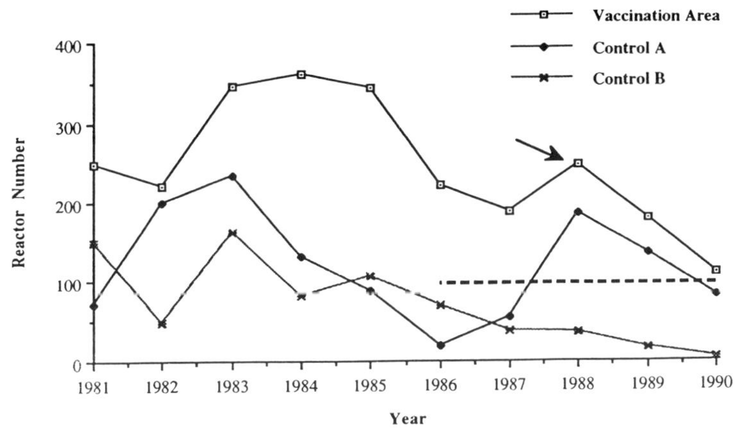

Fig. 2. Numbers of reactor cattle detected annually (1981-90) in the three experimental areas. Control A is a non-intervention area; badgers have been removed from Control B each year from 1986 onward (----); in the study area vaccination of badgers has been in progress since 1988 (➔).

RESULTS

Badger vaccination procedure

One of the principal objectives of this vaccination trial has been achieved. Badgers readily consume baits with chocolate-coated capsules. However, the amount of bait is critical as over-baiting can cause rejection. The aim is to keep the badger's curiosity aroused. Commonly, during the baiting period, increased rooting activity is seen around the setts. Yet, though the baits are placed in small amounts, often some peanuts are left in preference to chocolate.

Three people are required for a total of six weeks per season, i.e. spring and autumn, to complete the exercise. On each day, depending on road access to the setts, ten to fourteen setts are baited and a distance of approximately 5km (i.e. cumulative walks) is covered by each group.

Cattle data

Table 1 gives the total number of herds, animals, reactors and lesioned animals for each of the three areas for 1989 and 1990. Also included is the APT, i.e. number of reactors per thousand animal tests. The year 1989 is taken as a starting point for a cattle statistic baseline, that is the disease status in the year preceding vaccination. Individual herds are assigned, based on the number of reactors, to a seven-point scale extending from Range 1 (1 reactor) to Range 7 (31-50 reactors). See Table 2 for the number of reactors per range. For example, if multiple breakdowns decrease in size and number as a result of vaccination, a reduction would occur in the higher ranges, i.e. R4, R5, R6 and R7. Overall, a marginal improvement has been detected to date, but it is too early to comment further.

Table 1. The three experimental areas: results, 1989/90.

	No. of herds	**No. of animals**	**No. of reactors**	**No. of APT**	**No. of lesions**	**Year**
Vaccination	590	31,557	180	3.0	46	1989
area	595	32,068	111	1.8	41	1990
Control A	382	24,608	137	2.8	57	1989
	382	24,608	81	1.9	24	1990
Control B	152	6,884	16	1.8	3	1989
	152	6,884	4	0.44	1	1990

Table 2. Ranges 1-7, no. of reactors per range.

Range no.	**R1**	**R2**	**R3**	**R4**	**R5**	**R6**	**R7**
Reactors per range	1	2	3	4-10	11-20	21-30	31-50

DISCUSSION

From its inception, the aim of this project has been to vaccinate as many badgers as possible. In the pre-trial of 1988/9, vaccination was carried out during the summer months of June and July, specifically, to include badger cubs, mature enough by that time to eat the bait. However, for several reasons, beginning in 1990, this practice was altered. The current schedule involves baiting in early spring and late autumn of each year. The reasons are as follows: setts are more active, particularly in spring, and are easier to identify and more accessible; the absence of cattle on pasture in early spring, and of farm machinery in spring and autumn, creates less disturbance to badgers; the chocolate coating around the capsules remains intact in the absence of summer heat. Experience to date has shown that badger activity and vaccine uptake are significantly improved under this new schedule.

Interference with baits by non-target mammals was an initial worry, rats being the biggest offenders. Fortunately, this type of interference occurs only at a limited number of setts. Nocturnal observations to assess this aspect of unwanted interference showed that badgers had few competitors, not nearly enough to affect the trial (Stainsby, unpublished).

One nagging doubt remains. The percentage of badgers vaccinated is not known. In some families, dominant badgers may consume all of the bait and deprive the others. On successive days, baits are positioned on different tracks, and spaced further apart at 2m intervals rather than 1m. This routine is used to curb dominance. The process of biomarking could well provide the answer to the unknown percentage. Tetracycline, used as a biomarker, if placed in baits would on consumption be detected thereafter in bone by means of ultraviolet fluorescence microscopy (Capt 1981). A previous effort to have this process included in the trial failed.

In the study area, the TB incidence in infected cattle populations peaks every five to six years. Therefore, no definite conclusions can be drawn until at least five years have elapsed, i.e. 1993. However, indications of the effect, if any, of the vaccine may become evident sooner.

ACKNOWLEDGEMENTS

I am grateful to Dr Louis O'Reilly, Deputy Director, Veterinary Research Laboratory, Abbotstown, for valuable discussions during the development of criteria for use in selection of the experimental areas. I wish to thank Dr John Stanford, London University, for supplying the vaccine and for his keen interest in, and enthusiasm for, this project. My thanks to Professor Maire Mulcahy, Paddy Sleeman, Ph.D., Department of Zoology, University College, Cork, and members of the TB Investigation Unit, Dublin, for their guidance. Gratitude is also due to the following at the District Veterinary Office, Cork: H. N. Cummins, S.V.I., who has been very supportive; Tom Keohane, A.O., for his dedication and skilful fieldwork, ably assisted by Michael Creed, A.O.; and Aedan O'Meara, V.I., who helped in putting this report together. Finally, I am grateful to the personnel at ERAD headquarters for their encouragement.

REFERENCES

ANDERSON, R.M. and TREWHELLA, W. 1985 Population dynamics of the badger (*Meles meles*) and the epidemiology of bovine tuberculosis (*Mycobacterium bovis*). *Philosophical Transactions of the Royal Society* **310B**, 327-81.

BROCHIER, B., KIENY, M.P., COSTY, F. *et al.* 1991 Large-scale eradication of rabies using recombinant vaccinia-rabies vaccine. *Nature* **354**, 520-2.

CAPT, S. 1981 Köder Versuche beim Rotfuchs (*Vulpes vulpes,* L.). Lizentiatsarbeit, Universität Bern.

CHEESEMAN, C.L., JONES, G.W., GALLAGHER, J. and MALLINSON, P.J. 1981 The population structure, density and prevalence of tuberculosis (*Mycobacterium bovis*) in badgers (*Meles meles*) from four areas in south-west England. *Journal of Applied Ecology* **18**, 795-804.

CHEESEMAN, C.L., LITTLE, T.W.A., MALLINSON, P.J., PAGE, R.J.C., WILESMITH, J.W. and PRITCHARD, D.G. 1985 Population ecology and prevalence of tuberculosis in badgers in an area of Staffordshire. *Mammal Review* **15**, 125-35.

DOLAN, L. 1993 Badgers and bovine tuberculosis in Ireland: a review. In T. J. Hayden (ed.), *The badger,* 108-16. Dublin. Royal Irish Academy.

NEAL, E. 1986 *The natural history of badgers.* London. Croom Helm.

SMAL, C.M. 1993 The National Badger Survey: preliminary results for the Irish Republic. In T.J. Hayden (ed.), *The badger,* 9-22. Dublin. Royal Irish Academy.
STANFORD, J., STAINSBY, K. and MAHMOOD, K.H. 1993 Could vaccination be the answer to the badger tuberculosis problem? In T.J. Hayden (ed.), *The badger,* 174-80. Dublin. Royal Irish Academy.

In: Hayden, T.J. (ed.) 1993 *The badger*, 189-195. Dublin. Royal Irish Academy.

THE INTERACTIVE BEHAVIOUR OF CATTLE AND BADGERS WITH REFERENCE TO TRANSMISSION OF BOVINE TUBERCULOSIS

P.F.J. Benham*

Department of Pure and Applied Zoology, University of Reading, U.K.

*** Present address: Primrose Farm, Felindre, Brecon, Powys, LD3 OST, Wales**

ABSTRACT

Behavioural studies explain why the rate of transmission of *Mycobacterium bovis* from badgers to cattle is surprisingly low and in places non-existent despite the presence of infected badgers. Night observations on badgers and cattle indicated that direct transmission via inhalation by cattle of an expired aerosol from the badger is unlikely. Chance of direct contact was minimal when cattle were congregated in groups. Most cattle avoided grass contaminated with badger faeces and urine, but when grass was scarce a few unselective individuals inhaled more and grazed closer to the badger products. These individuals are at risk when herbage is scarce. Badger setts in pastures pose a serious risk as cattle performed many activities at their entrances. Wild badgers were attracted to and ate cattle foods throughout the year. Badger passage to desirable resources was successfully prevented with electrified flexinet. Management recommendations are produced that will minimise the risk of disease transmission from badgers to cattle.

INTRODUCTION

A small proportion of the badger population in some areas of the United Kingdom is infected with *Mycobacterium bovis* (Dunnet *et al.* 1986) and in certain situations transmission of tuberculosis (TB) from badgers to cattle could occur. However, the frequency and rate of transmission has been found to be surprisingly low even in highly confined artificial conditions (Little *et al.* 1982). There are examples in the natural situation in which transmission between species does not occur. Surprisingly, rabbits, although very susceptible, do not contract TB in the wild. Cobbett (1917) suggested a behavioural explanation. Although in Switzerland some badgers are infected with TB, they do not appear to pose a threat to cattle (O'Connor and O'Malley 1989). There are known examples in the UK where, considering the levels of TB in the badger population, the incidence of TB in the surrounding cattle was non-existent or much lower than expected (e.g. Folkington in Sussex and Woodchester Park during the period 1981-7). In these situations it is likely that the badgers and the cattle interact in such a way as to prevent transmission occurring. Studies at Reading University on the interactive behaviour of cattle and badgers identified the most likely situations in which transmission of *M. bovis* from badgers to cattle could occur (Benham 1984). Utilising this information, management practices can be devised to keep cattle at a distance from badgers and their products, and this will minimise transmission from badgers to cattle. The most likely mode of transmission between humans is considered to be inhalation of expired droplets. Humans working with infective cattle are most at risk through direct contact (Francis 1958). Work in New Zealand (J. Coleman, pers. comm.) suggests that direct transmission of TB from possums to cattle is likely to be an important factor involved in disease spread. When possums are threatened, they stand their ground and from an upright position scream and spit at the intruder.

RESULTS

Chance of transmission by direct contact

The risk of transmission of *M. bovis* to cattle via air expired by badgers is very low. All activities of wild badgers in cattle fields at three different sites were observed at night for 359 hours (Benham and Broom 1989). Both whilst badgers were moving through pastures and when foraging within pastures they were clearly aware of the positions and actions of nearby cattle and preferred to remain at least 10-15m from the cattle. Some cattle would move towards badgers, particularly if they were performing an interesting activity such as collecting bedding, but badgers fled rapidly from every approach. They responded rapidly to cattle movements, and visual monitoring could be observed since the badgers regularly looked towards nearby cattle. A video-recording made of cattle and badgers in situations of potential contact clearly demonstrates the way in which badgers visually monitor cattle and avoid close contact with them. In all potential encounters, badgers were able to keep at least 2-3m from approaching cattle. In an experiment at the Grassland Research Institute, on five out of fifteen nights cattle were confined in a small field which was next to the badgers' main sett and through which the badgers always travelled. On the nights when the cattle were present, the badgers altered their routes to pass around the field containing the cattle. During these observations badgers clearly avoided compact groups of cattle more than individuals. Also, standing cattle were considered a greater threat than those lying.

These results were supported by studies in a 0.5ha enclosure with an artificial sett and containing cattle managed either on set-stocked or rotational grazing systems. All activities of pairs of wild-caught badgers were observed for the duration of 20 nights when cattle were present. Badgers in the artificial sett delayed entering the enclosure if cattle were within 15m of the entrance. In the first two trials the cattle were confined to one of eight areas (one area per night) on a rotational (strip-grazed) system. The badgers were able to travel freely around the enclosure. However, they generally retreated from or travelled around the area occupied by the cattle. In the two trials they spent 10 and 23 times less time in the cattle areas than would be expected by chance. During the third trial, the cattle were set-stocked and were more dispersed and hence occupied more sections of the enclosure. During the set-stocked trial, the badgers carried out 41.2% of their normal behaviour in the same areas as cattle. Clearly, the badgers were unable to keep a large distance from individual cattle dotted around, and adapted to a closer proximity to cattle than on the rotational grazing system. Other studies on wild badger populations showed that badgers were rarely present in pastures during dry conditions and their presence was variable at other times. A preference was found for foraging in long clover rather than shorter grass (Benham and Broom 1989).

Chance of transmission whilst grazing

The responses of grazing cattle to badger latrines in their pastures gave no indication that cattle were attracted towards badger latrines from a distance in order to investigate them. In fact, the duration of time spent within, and time spent grazing, was less in latrine areas than in control areas.

The responses of individual animals were assessed by presenting turves of grass, either clean or containing badger urine or faeces on one of four areas, to housed cattle. There was a strong avoidance of badger faeces even when weathered for up to 28 days. However, one out of 30 cows did eat grass from the faecal-contaminated area. Similarly, with fresh badger urine, most cows avoided the contaminated area. The cows exhibited a range of selectivity in which very selective animals rejected a whole turf. With decreasing selectivity, animals took bites closer to the contaminated area and the unselective cow ate all contaminated grass. Similarly, the response to grass contaminated with urine up to 14 days before presentation was generally avoidance, except that cows became slightly less selective as the time since contamination increased (Benham and Broom 1991).

Studies were also performed on grazing cattle by recording responses to 1m^2 plots which were either untreated controls or had been contaminated with badger urine or faeces. In a herd of 150 cows, only one individual ever grazed on faecal-treated plots. She also grazed most in urine-contaminated plots. Four other cows, on occasions, grazed from urine plots. Most of the grazing of contaminated plots occurred at later stages of progressive defoliation when grass was scarce. When grass was abundant, the contaminated areas were totally avoided. When herbage was scarce, the selective individuals still avoided from a distance, but moderately unselective animals sometimes performed more olfactory investigation and unselective animals grazed contaminated grass. Thus when herbage is short there will be a greater risk that some animals will take in *M. bovis* by inhalation or ingestion. The unselective cow also ate grass contaminated with bird droppings or cow dung that had been rejected by other cattle (Benham and Broom 1991).

Cattle age and breed susceptibility to TB

Data were collected from farms on which breakdowns had occurred and had been attributed to badger origin. Within each cattle group the age of the reactors was compared with the age of other group members. It was demonstrated that the risk of infection progressively increases with age of the cattle. This could be related to the fact that older cattle are less selective in their grazing behaviour (Hafez and Bouissou 1975). This result is consistent with the hypothesis that less selective cattle are at a greater risk of taking in an infective dose of *M. bovis*. With limited data, a possible trend showed Friesians to be more susceptible than other breeds. Modern management and breeding, particularly of dairy cows, will tend to discourage selectivity and encourage opportunist unselective individuals.

Transmission at badger setts

Observations on cattle at badger setts present in their pastures showed that cattle may spend considerable time at sett entrances. Head-muddying is performed and this elicits social interest. Many cattle may cluster around, put their noses down the holes and lick the ground. They will also lick soil off the faces of other cattle. Badgers were recorded regularly marking at sett entrances in pasture even when not resident in them. This, together with infected bedding and other contamination, creates a high-risk situation.

Transmission at food sources

Preferences for cattle feeds of different types were assessed in the wild badger population at Woodchester Park over the period of a year. The badgers avidly ate rolled barley and dairy nuts throughout the year. Up to 220 minutes per night were spent at the barley feeder. When other food was scarce they also ate maize silage, rumevite and molasine meal. A sick badger, perhaps unable to forage efficiently, might find an 'easy' food source of this nature very tempting. Therefore, supplementary feeding of cattle at pasture could increase the chance of direct contact occurring and also cause contamination of food troughs.

Transmission in farm buildings

There have been a number of sightings of badgers entering farm buildings in order to eat cattle food (Edwards 1986; Kruuk 1989; Sleeman and Mulcahy 1993). Sick badgers have been observed living in farm buildings (Cheeseman and Mallinson 1981) and have also been found dead in them. Sick badgers will be encouraged to take up residence and remain in farm buildings if they have access to attractive food such as corn or cattle food. This is a high-risk situation and transmission could occur either by contamination of food or through direct contact in unusual circumstances.

Badger exclusion using electrified fencing

An electric wire was erected around the enclosure approximately 5" above the ground to prevent the local wild badgers fighting through the fence with the captive badgers. After first contact a badger was seen fleeing rapidly across the field. However, after two weeks the very uneven ground beneath the wire enabled the badgers to learn to push under the wire while protected by their thick coats. Electrified flexinet was then erected around the enclosure (1-3m from the perimeter fence). Badgers regularly patrolled around the flexinet and scraped at vegetation and ground outside it but they never broke through the flexinet over a five-month period, even though the fence was switched off for two weeks. Badger exclusion from areas surrounded by flexinet was observed at GRI. For two weeks a clover field provided a premium foraging site, with up to 50 badger-hours spent foraging per night at locations well distributed through the field. After sheep were placed in three-quarters of the field, surrounded by flexinet, the badgers never entered this part and only foraged in the remaining quarter.

MANAGEMENT RECOMMENDATIONS TO MINIMISE THE RISK OF TRANSMISSION

1. Prevent cattle access to badger setts

On the twelve farms visited following herd breakdowns, cattle often had access to badger setts. Since cattle are very active at sett entrances, this poses a considerable risk of transmission through cattle contact with infected bedding, urine, wound discharge, or even rising expired warm air. Badger setts remain generally in the same position, so they can be fenced off to prevent cattle access without great difficulty.

2. Prevent badger access to palatable feeds in farm buildings

Any food around farm buildings such as present in animal food and grain stores should be made badger-proof since an attractive source would encourage sick badgers to remain in buildings. Maize silage clamps could also pose a serious risk.

3. Feed cattle at pasture in badger-proof troughs

Shallow troughs should be held more than 80cm above the ground by a rectangular open metal framework which has no cross-struts or diagonal bars which badgers could climb. Milking parlours and milking bails in fields pose a risk since badgers may contaminate feed troughs.

4. Exclude badgers using electrified wires or flexinet

Badger access to high-risk sites such as farm buildings or maize silage clamps should be prevented using electrified fencing. Badger access to high-risk grazing areas could be prevented particularly in spring and autumn. One herd-owner in Northern Ireland has surrounded his farm with flexinet (G. Russel, pers. comm.).

5. Adopt a 24-hour rotational grazing system

This will confine the cattle to a small area and will maximise the badger avoidance behaviour. Within this grazing system there will be less chance of transmission occurring by direct aerosol contact. Also the chance of taking in an infective dose of *M. bovis* from environmental contamination will be less than in a set-stocked system. This system will prevent familiarisation occurring between particular groups of badgers and cattle. The effect of this system will be maximised when the cattle are moved to their fresh daily herbage in the afternoon, since any badger products in the new grass will have undergone considerable weathering since deposition. This system will also be most effective when used in spring and autumn when cattle perform more night grazing and badgers are more active in pastures. The greatest beneficial effect will be found with adult dairy cows, since beef cattle and young stock tend to perform less night grazing and lie down in more congregated groups (Benham 1984; Benham and Broom 1991).

6. Do not graze cattle hard

As herbage becomes scarce, there is a greater risk of some unselective animals receiving an infective dose of *M. bovis* through inhalation or ingestion followed by eructation. If cattle are kept on a set-stocked system then the stocking rate should be low. If they are rotated around herbage areas, they should be moved frequently and ideally followed by sheep which could safely eat the contaminated herbage rejected by cattle.

7. Provide an alternative foraging area for badgers

If the badger preference for clover is universal, then badgers would be attracted to clover fields away from cattle grazing areas and the clover could be used for hay-making or seed.

8. General cattle husbandry

A general aim should be to produce strong, healthy cattle that can resist a small disease challenge. Poorly fed cattle and those kept in bad conditions are likely to be more susceptible to disease. In addition, many modern intensive breeding and husbandry practices tend to reduce vigour and vitality. Examples are social stress (Benham 1984), corticosteroid levels, over-feeding concentrates to ruminants (Webster 1987) and overuse of drugs and antibiotics.

CONCLUSIONS

Behavioural studies on badger/cattle interactions have identified the most likely ways that transmission of TB would occur from badgers to cattle. This provides an understanding of why the rate of transmission to cattle is surprisingly low, and non-existent in some situations. There is no chance of eradication of TB using control measures (Dunnet *et al.* 1986). Thus, the only realistic aim now can be to minimise transmission, The management recommendations will minimise the risk of transmission. The immune response to *M. bovis* in the badger population should be encouraged. A long-term policy can only be successful if it is based on these sound ecological and behavioural criteria.

REFERENCES

BENHAM, P.F.J. 1984 Social organisation in groups of cattle and the interrelationships between social and grazing behaviours under different management systems. Unpublished Ph.D. thesis, University of Reading.

BENHAM, P.F.J. and BROOM, D.M. 1989 Interactions between cattle and badgers at pasture with reference to bovine tuberculosis transmission. *British Veterinary Journal* **145**, 226-41.

BENHAM, P.F.J. and BROOM, D.M. 1991 Responses of dairy cows to badger urine and faeces in pasture with reference to bovine tuberculosis transmission. *British Veterinary Journal* **147**, 517-32.

CHEESEMAN, C.L. and MALLINSON, P.J. 1981 Behaviour of badgers infected with bovine tuberculosis. *Journal of Zoology* **194**, 285-9.

COBBETT, L. 1917 *The causes of tuberculosis.* London. Cambridge University Press.

DUNNET, G.M., JONES, D.M. and McINERNEY, J.P. 1986 *Badgers and bovine tuberculosis.* London. HMSO.

EDWARDS, M. 1986 *Badger valley.* London. Michael Joseph Ltd.

FRANCIS, J. 1958 *Tuberculosis in animals and man.* London. Cassel and Company Ltd.
HAFEZ, E.S.E. and BOUISSOU, M.F. 1975 The behaviour of cattle. In E.S.E. Hafez (ed.)*The behaviour of domestic animals* 203-245. (3rd edn). London. Bailliere Tindall.
KRUUK, H. 1989 *The social badger.* Oxford University Press.
LITTLE, T.W.A., NAYLER, P.F. and WILESMITH, J.W. 1982 Laboratory studies of *M. bovis* infection in badgers and cattle. *Veterinary Record* **111**, 550-7.
O' CONNOR, R. and O'MALLEY, E. 1989 *Badgers and bovine tuberculosis in Ireland.* Dublin. ERAD/ERSI.
SLEEMAN, D.P. and MULCAHY, M.F. 1993 Behaviour of Irish badgers in relation to bovine tuberculosis. In T.J. Hayden (ed.), *The badger,* 154-65. Dublin. Royal Irish Academy.
WEBSTER, J. 1987 Feeding problems and metabolic disease. In *Understanding the dairy cow.* 175-203. Oxford. Blackwell Scientific Publications.

In: Hayden, T.J. (ed.), 1993 *The badger*, 196-211. Dublin. Royal Irish Academy.

THE BADGER: EPIDEMIOLOGICAL DARKNESS IS PLIGHT ENOUGH

T. J. Hayden

Mammal Research Group, Department of Zoology, University College, Dublin

OVERVIEW

A number of important conclusions may be drawn from the papers included in this volume and from the discussions stimulated by them.

1. Although the badger is currently under study by a number of research groups throughout Europe, it has been inadequately studied in Ireland until relatively recently. It is to be anticipated that further investment in fundamental research will contribute substantially to our understanding of what is an unusual mustelid carnivore.
2. Despite misleading comments, the role of the badger in the aetiology of tuberculosis in bovines is still not quantified, partly due to the intractable scale of field experimentation required and the lack of even a simple model of the pathways through which *Mycobacterium bovis* circulates through wildlife and domestic livestock.
3. There is an increasing awareness that the conflict, irrespective of its basis, between conservation imperatives and pragmatic agricultural interests, in so far as the badger is concerned, is not so much a debate about priorities but rather an attempt to evaluate conflicting viewpoints without appropriate data. This is further exacerbated by the urgency with which the problem of tuberculosis in bovines must be controlled.

An adequate understanding of what is a complex multifactorial process will require a multidisciplinary approach if progress is to be made.

BIOLOGY OF THE BADGER

The sociality of badgers

The studies reported here raise a number of interesting questions relating to the nature of sociality and its implications for control of bovine TB. In general, badgers live in groups occupying a communal area, which is defended by a system of boundary latrines at which various scents are deposited (Kruuk 1989; 1993). However, it is becoming increasingly apparent that the badger exhibits a wide range of social systems, although the factors responsible for the variants are less clear. Notions of what is the typical condition for badgers are to some extent influenced by the frequency with which particular social systems are studied. It is therefore of interest to extend investigation of social systems in various habitats or other circumstances. It is becoming increasingly clear that the habitats selected for the construction of setts in Ireland differ from those in Great Britain (Cresswell *et al.* 1990). Three levels of analysis, local, regional and national, all indicate that hedgerows are of great importance and contain about 50% of setts (O'Corry-Crowe *et al.* 1993; Eves 1993; Smal 1993). This is an unusual situation in that the disposition of setts is probably a derived feature due to the extent of deforestation which has taken place in Ireland in historical times. Nevertheless, it appears from the preliminary accounts of the survey of badger setts that the population of badgers in Ireland is probably about as large as that of Great Britain (Smal 1993; Feore and Montgomery 1993), although the routes by which badgers colonised the island are far from clear (Lynch *et al.* 1993).

Territories are normally identified using the bait-marking approach as first used by Kruuk (1978; see also Cheeseman, Mallinson *et al.* 1993). This method reveals the boundary latrines and a minimum polygon is constructed to represent the territory. Many of the sides of these polygons coincide with well-used badger paths. These territories are believed to represent areas containing a resource which the occupants are prepared to defend. This does not imply that the occupants are constrained within these territories; indeed, animals might be expected to cross the boundaries if the opportunity to exploit safely the resources of their neighbours presents itself. Thus radio-tracking studies will reveal overlaps in home ranges of individuals in adjacent social groups as transgressions are progressively recorded. Radio-tracking studies may underestimate the extent of territorial defence in the absence of evidence from bait-marking. Thus the report that the badgers in Offaly are highly territorial (bait-marking; O'Corry-Crowe *et al.* 1993) while those from Cork exist in less well-defined social groups (radio-tracking and bait-marking; Sleeman and Mulcahy 1993) may be due to a different methodological emphasis as well as different degrees of persecution. Clearly there is still much to be learned from an intensive study of the behaviour of individual badgers in relation to the so-called activities of the group of which, de facto, they are a part.

The Resource Dispersion Hypothesis has been a major stimulus to research on the social organisation of badger populations (Macdonald 1983; Kruuk and Macdonald 1985; Carr and Macdonald 1986). On this view the disposition of earthworm-rich areas was considered to determine the configuration of the minimally viable territories, which, however, were often capable of supporting additional badgers. Additionally, badgers were regarded as being essentially contractionist, rates of dispersal were low. However, Roper (1993) has strongly argued that the dispersion of setts or substrates suitable for their excavation could equally be the key resource determining sociality. Furthermore, recolonising badgers moving into cleared areas in Gloucestershire behaved in rather a more expansionist mode and occupied large home ranges before restricting their activities to a more normal-sized territory (Cheeseman, Mallinson *et al.* 1993). Whether such a phenomenon is occurring or will occur in the much larger clearance area in County Offaly remains to be seen. Nevertheless, analysis of recolonisation of areas by badgers will allow opportunities to examine some hypotheses in relation to the evolution and development of sociality in badgers (Woodroffe and Macdonald 1992); but whether the development of social groups in a depleted region of suitable habitat recapitulates the phylogeny of sociality is quite another matter. Furthermore, the fitness implications of sociality for individual badgers in a group have yet to be quantified. For example, many females in groups fail to rear cubs when it seems there is little impediment to successful reproduction by a female occupying a territory as a member of a breeding pair of adult badgers (Whelan and Hayden 1993). The developments in technology to address the genetic structure of badger populations hold out great promise for advances in our understanding of global and local evolutionary processes (Evans *et al.* 1989; Lynch *et al.* 1993). Application of paternity and maternity tests to populations which have been the subject of long-term study will be of great interest (Cheeseman, Wilesmith *et al.* 1987; Harris and Cresswell 1987; Sleeman and Mulcahy 1993). It seems that we are about to witness a paradigm shift in our understanding of the sociality/antisociality of badgers.

BADGERS AND BOVINE TUBERCULOSIS: THE EVIDENCE

It is appropriate to consider the evidence for a role of the badger in the transmission of TB to cattle as presented formally or discussed in this seminar and the various interpretations placed on such data. Evidence brought to bear in this debate, so far, is mainly of 7 types:

(i) the demonstration that transmission of TB from badgers to cattle can occur,
(ii) the fact that TB is widespread in badger populations,
(iii) the coincidence of infected badgers and infected herds,
(iv) identity of isolates of *M. bovis* from badgers and cattle in the same region,
(v) field indications of high-risk individual badgers (moribund or dead infectious individuals) or presence of badger products likely to be infectious,
(vi) the relationship between removal of badgers and subsequent progress of TB in the local cattle, and
(vii) the relationship between vaccination of badgers and the course of TB in local cattle.

Proven transmission

In 1982 successful transmission of tuberculosis from badgers to cattle was reported (Little *et al.* 1982). This paper seems destined to become a citation classic since it has been pressed into service by both the advocates of a major and a minor role for the badger in the aetiology of TB in bovines. For example, the Dunnett report cites this paper in support of the view that the badger was the primary source of tuberculosis in cattle (Dunnett *et al.* 1986) and many participants in this seminar considered the fact that badgers infect cattle to be beyond question. But Benham (1993) draws attention to the fact that Little *et al.* (1982) considered that the risk posed by badgers, even when confined together with cattle, was low. Quantification of the risk is the major epidemiological black spot remaining to be solved. Although no new definitive evidence was presented in this seminar the final outcome of the work reported by Eves (1993) and McCarthy (1993) is awaited with interest.

Tuberculosis in badgers

Tuberculosis is clearly endemic in badgers in Ireland (Dolan 1993). About 17% of badgers removed by the Department of Agriculture and Food revealed lesions at necropsy, but the real rate of infection is more likely to be about 25% (Fagan 1993). Extrapolation of these rates to the whole country is likely to produce an overestimate, since these figures are derived from animals taken from areas where badgers are suspected to be the cause of the outbreak under investigation in cattle. It is not clear if there are regional differences in the incidence of TB in badgers since the sample sizes for badgers obtained other than in culls is low. The generally higher incidence of TB in badgers, when compared to cattle, is due, among other things, to ease of transmission between members of the same clan, together with the longer life expectancy of an infected badger (Cheeseman, Cresswell *et al.* 1988). Infected cattle are removed from herds as they are identified in annual herd tests. Most of our information on rates of infection in badger populations is derived from animals post-mortem. What are the implications of a diagnostic test which could be applied *in vivo*? An ELISA-type assay which is currently undergoing field trials in Britain was described in this seminar (Nolan and Goodger 1993). This appears to provide the possibility of rapidly evaluating the TB status of an individual badger, but it is less clear how this test might be incorporated into a scheme whereby infected badgers only were to be culled. There are major difficulties with such a scheme, among which are (1) the likelihood, in Ireland at least, that there are some infected badgers in all social groups, (2) the cost and labour-intensive nature of a comprehensive and humane trapping effort, and (3) the uncertainty as to whether the exercise would be beneficial on a wide scale. A case, however, could be made for a field trial of this type.

Correlations

Data presented by Dolan (1993) and Smal (1993) allow a preliminary examination of the relationship between the incidence of TB in badgers, the incidence of TB in cattle, and the density of badgers on a regional basis. It is clear that there is not a simple and obvious relationship between these variables. Two points, however, need to be emphasised. Firstly, questions may be posed about the appropriateness of comparison of the data across these two papers, but it is raised here to make explicit that these are the types of data on which intuitive decisions have hitherto been based. Secondly, it is not immediately clear which is the dependent variable. Little attention has been paid to the possibility that TB in badgers, at least on the wider scale, is to some extent dependent on the incidence in bovines, as has been argued. Furthermore, it has so far proved impossible to demonstrate a significant correlation between badger density (assessed as setts per km^2) and disease incidence across counties in the UK (Cheeseman, Wilesmith *et al.* 1989). However, there was a relationship between 7-year cyclicity of TB and the population density of badgers in an intensively studied population in Gloucestershire (Cheeseman, Wilesmith *et al.* 1989). Overall this does not exclude the possibility that there may exist a threshold density of badgers for maintenance of infection across a wider geographical area, as was modelled by Anderson and Trewhella (1985) and assumed by McCarthy (1993).

Genotypes of M. bovis *from badgers and cattle*

It has long been assumed that the organism which infects cattle is the same as that isolated from badgers although the taxonomy of the tubercule bacilli is rather difficult to resolve. A study of 150 cultures of *M. bovis* from cattle, deer and badgers in Ireland identified two strains but failed to detect sufficient biochemical differences which could be used to subdivide isolates for epidemiological studies (Grange *et al.* 1990).

Furthermore, this study also reported that preliminary attempts to identify heterogeneity by restriction endonuclease cleavage of extracted DNA or to demonstrate restriction fragment length polymorphism were unsuccessful. These findings have been cited in this seminar by Dolan (1993) and were advanced by some participants as evidence that badgers were infecting cattle. This needs careful consideration because the fact that badgers and cattle in a particular area harbour the same strain still does not clarify the polarity of the cross-infection. Secondly, these results indicate either that *M. bovis* in Ireland is particularly homogeneous, unlike the bacillus in New Zealand (deLisle *et al.* 1990), or that the genetic variation in *M. bovis* is too subtle for the techniques so far employed (Grange *et al.* 1990). If indeed *M. bovis* in Ireland proves to be rather homogeneous this may still not be due entirely to the frequency of bidirectional cross-infection between cattle and badgers but could as likely be due to the highly mobile nature of cattle in this country. It seems more likely that there is genetic variation in *M. bovis* yet to be revealed. If so, it might be possible to discriminate between outbreaks of TB in cattle where the genotype of *M. bovis* is shared with those in sympatric badgers (suspected badger involvement) and cases where the genotypes of the bacilli from cattle and sympatric badgers are different (badger involvement highly unlikely). This is a fruitful avenue of research to pursue, particularly if polymerase chain reaction techniques are applied to the problem (Barry *et al.* 1990).

Field evidence of transmission of TB from badgers to cattle: tempo and mode

When an outbreak of TB occurs in a herd of cattle the acute response involves removal of reactor animals, investigation of the factors which may have contributed to the outbreak, and recommendation of remedial action. Examples of this process have been given by Flanagan (1993), who pointed out that the badger is not treated as the prime suspect unless there is strong circumstantial evidence and there are no other factors identified which represent a more likely cause. The difficulties faced by field officers are revealed when it is understood that of the broad categories of sources of infection outlined by Flanagan (1993) none can be quantified and the transmission coefficients of all or any of the pathways are quite unknown. Nevertheless it must be emphasised that, whatever the basis of epidemiological decisions, in only 14% of 3975 epidemiology reports was wildlife (mainly the badger) suggested as the proximate source of the infection (Griffin and Hahesy 1992). Thus a great deal of heat might be taken out of the debate if these data were more generally known. The acute response to a strong suspicion of involvement of badgers currently is to remove social groups which (presumably) contain infected badgers and whose location is such that these badgers forage over the farm in question. This involves executive action by the Department of Agriculture and Food which may be carried out expeditiously. Further action such as fencing is recommended but is at the discretion of the farmer. Removal of badgers requires a licence from the Wildlife Service. The conditions under which such a licence is issued are discussed in this seminar by Warner (1993). The welfare implications of the control operation must now be considered. The Wildlife Service lays down conditions as to the type of trapping equipment to be used and the method by which badgers should be killed. However, of major concern is the lack of a close season. Experience over a decade, mainly in Gloucestershire, has indicated that complete removal of social groups is possible, but that a substantial number of traps/snares should be used and they should be set for ten consecutive nights (Cheeseman and Mallinson 1979). A similar conclusion was reached by O'Corry-Crowe *et al.* (1993), who concluded from an analysis of the trapping effort returns from episodic trapping in east Offaly that about 500 trap-nights should be applied to each social group and that the trapping effort should be applied over at least ten consecutive nights. This should ensure that the objective of removal of the social group is achieved. However, since most cubs are born between the end of January and the end of February and at least some of the females may still be nursing cubs in June (Whelan and Hayden 1993), there are cogent welfare arguments for restricting trapping to the second half of the year.

Possible routes for transmission of TB from badgers to cattle involve contact between cattle and badgers or their products. Reaction of cattle to badgers, as reported in this seminar, varies from active interest (Sleeman and Mulcahy 1993) to mutual indifference (Benham 1993; Brown *et al.* 1993). Reaction of badgers to cattle, on the other hand, has ranged from active avoidance to mutual indifference. It seems likely that the extent of interaction depends on the previous experience of an individual of one species in encounters with members of the other. Investigative behaviour of cattle towards dead or moribund badgers has often been reported by veterinarians in Ireland, but no systematic records are available, so the extent of this phenomenon is not clear. This is unfortunate since if these dead or dying badgers are tuberculous they must constitute a high-risk factor. A more widespread source of infection is represented by badger products, faeces, urine or sputum. It has been considered that cattle may become infected by ingestion or inhalation of *M. bovis* while grazing contaminated pasture or investigating other sites used extensively by badgers such as latrines or the environs of setts (Muirhead *et al.* 1974; Benham and Broom 1989). Badgers forage extensively over pastures in search of their staple food item, earthworms (Kruuk 1978). Therefore the extent of contamination of pastures by urine and faeces and the reaction of cattle to such contaminated patches are of great interest in devising management strategies to minimise the risk of infection. Although it has been shown, in choice tests, that cattle normally avoid eating contaminated herbage, nevertheless some animals are less discriminatory (Benham and Broom 1991). Furthermore, it has been demonstrated that most but not all cattle will avoid areas of pasture artificially contaminated with badger faeces or urine, but that in the later stages of defoliation in heavily grazed pastures cattle more frequently investigate or graze contaminated areas (Benham and Broom 1991). Accordingly a comprehensive package of management procedures which should decrease the likelihood of transmission of TB to cattle from contaminated pastures has been described by Benham (1993). It was of course still difficult to evaluate the risk of contamination, but it seems likely that contamination of pasture by faeces would be of lesser importance since badgers mainly defecate in aggregations of dung pits, termed latrines, which serve as indicators of territorial boundaries to neighbouring social groups and are of great strategic significance to badgers. Such latrines are often found in association with topographical features such as field boundaries (see for example Cheeseman, Mallinson *et al.* 1993). This is particularly so in Offaly (O'Corry-Crowe *et al.* 1993). The urinary route must now be considered. Although infected badgers frequently exhibit lesions in the kidneys (Fagan 1993; Dolan 1993) and the microanatomy of the kidney is conducive to the release of bacteria to the urine (Elgadi and Hayden 1993), which may become extremely heavily contaminated, to the extent of up to 300,000 bacilli mL^{-1} (Gallagher and Horwill 1977), nevertheless the pattern of deposition of urine within the home range of a badger has been little studied. This problem has inspired the development and application of novel and ingenious techniques to examine the fine structure of habitat use by individual badgers (Brown *et al.* 1993). Already this approach is bearing fruit. It is now becoming apparent that badgers do not urinate randomly over pastures as has hitherto been suspected. Rather, many badgers, particularly adults and yearlings, deposit a high proportion (97%) of their urine at latrines or on paths. Cubs, on the other hand, urinated in a random pattern. These findings are of great biological and agricultural interest. Firstly they suggest that urinary components may play a more important role in scent-marking territorial boundaries between social groups, and secondly that the logistical problems of denying access of cattle to contaminated areas are more amenable to solution (Brown *et al.* 1993). For example, in east Offaly over 90% of boundary latrines were less than 5m from a natural boundary and many were within the canopy of the hedgerow vegetation, and 70% of the pathways which interconnect these latrines were within 2m of the field edges as defined by the graze line (O'Corry-Crowe and Hayden, in prep.). Thus it is now apparent that the recommendation of double fencing to minimise lateral spread between cattle could, by taking field signs of badgers into account, produce additional benefits at no extra cost by reducing the possibility of contact between cattle and contaminated patches. Thus the case for adequate fencing is strengthened.

The attitudes of farmers to the perceived risk by badgers are complex. On the one hand the circumstantial evidence implicating the badger in a proportion (about 14%) of herd breakdowns is considered sufficiently compelling to warrant badger control (Griffin and Hahesy 1992), yet in a study site in east Offaly of 55 setts of various categories which were used during a two-year period, 35 (63%) were accessible to cattle and those which were not were so because of their location (O'Corry-Crowe *et al.* 1993). Thus of all the setts in the area none had been fenced off to exclude cattle. Qualitatively similar observations have been reported from the study site in County Cork (Sleeman and Mulcahy 1993). Thus the response of farmers to advice that they carry out what might appear to be intuitively obvious seems to indicate that the perceived benefits do not warrant the costs involved. Perhaps a cheaper solution to the problem of reducing contact between badgers and cattle is being practised, i.e. destruction of the badgers or their setts. This is illegal and seems unlikely to be happening on a grand scale. While the evidence of interference with setts (15%) in the national survey is clearly a matter of concern it does not support the view that the degree of interference is related to the extent of the tuberculosis problem in bovines (O'Connor and O'Malley 1989; Smal 1993). There certainly is intervention by farmers who are coping with an outbreak of TB in their cattle (Sleeman and Mulcahy 1993) but it is not widespread even in so-called black spot areas (O'Corry-Crowe *et al.* 1993). Therefore, although there can scarcely be a farmer who is unaware of the debate on the role of the badger in the bovine TB problem, prophylactic destruction of badgers is not a common practice. It is likely that the current extent of interference with badger setts is more to do with other pursuits, since in about 50% of cases of disturbance recorded during the national badger census evidence of digging was noted (Smal 1993). This is more indicative of efforts to obtain live badgers for illegal purposes than to reduce rapidly the population on the farm.

Tuberculosis in badgers

It is implicit in most discussions of the aetiology of TB in bovines that recurrent seeding of infection by badgers is a major cause of the persistence of the disease in the national cattle herd and that TB in badgers is self-sustaining. Nevertheless the progress of the disease through badger populations is far from easily understood. Tuberculosis in badgers may exhibit cyclical changes in prevalence, which may or may not run in phase with changes in the pattern of disease in cattle (Cheeseman, Wilesmith *et al.* 1989). Thus the view that TB is sustained in cattle *mainly* by cross-infection from badgers is a hypothesis that remains to be tested.

Information on the microgeographical variation in the extent of infection in badgers is currently rather sparse. It has been suggested that reduction in the population density of badgers would be expected to reduce the incidence of TB in badgers, which, for example, in the case of east Offaly is currently running at about 25% (Eves 1993). There is little evidence of a density-dependent degree of infection within social groups notwithstanding the high frequency of interactions between their members (Cheeseman, Wilesmith *et al.* 1989). However, the fact that the relationship between density and extent of infection is non-linear does not exclude the possibility that there may be a threshold density of badgers required to maintain the disease within them. To date, culling of badgers has been carried out in areas with concurrent high incidence of disease in both cattle and badgers. An apparent acute effect of culling might be generated if the more highly infected social groups were replaced by founder immigrants from areas with a lower rate of infection. Follow-up culls would then reveal a lower density of badgers and a lower rate of infection, but to infer a causal relationship would not be appropriate in the absence of further information on the distribution of tuberculous badgers in the area.

East Offaly: background

The incidence of bovine tuberculosis in cattle in County Offaly is among the highest in Ireland (Eves 1993). Since 1978 between 5.0% and 11.3% of herds contained reactor animals. Over the same period the reactor rate expressed as a percentage of the total population ranged from 0.27% to 0.75% (O'Connor and O'Malley 1989). Outbreaks of TB are not uniformly distributed throughout the county and in particular areas the incidence is substantially higher. For example, since 1980 in the Killeigh, Rahan and Croghan areas the reactor herds varied between 3.7% and 40.6%, between 7.6% and 41.8%, and from 11.1% to 36.1% respectively. In the mid-1980s badgers were removed from these three areas and in each area the proportion of reactor herds and the incidence of reactor animals declined. In one case the proportion of reactor animals decreased within 2 years to values which had not been recorded for 3 to 6 years prior to removal of badgers (O'Connor and O'Malley 1989). Similar observations from areas outside Offaly and individual case studies where there was circumstantial evidence to indicate a possible involvement of badgers provided the impetus to examine the effect of badger removal over a wider area. Thus the east Offaly project was undertaken to assess the extent to which the prevailing high incidence of TB in cattle might be influenced by the controlled removal of badgers in comparison with another designated area from which badgers would not be removed. The project began in November 1988. From 1988 to 1991 bovine TB decreased by 52% in the project area and increased by 33% in the surrounding area.

The overall pattern of variation in the incidence of TB in cattle may be produced by several types of variation. These include irregular fluctuations, seasonal changes in incidence, cyclical changes in annual rates, all with different periodicities, superimposed on longer-term trends. The incidence of TB in bovines is cyclical and up to 30% of the overall rate may be accounted for by these fluctuations which recur with a periodicity of about three years (Crilly, unpublished, in Downey 1990). Nevertheless it is possible, in principle, firstly to identify the position in the cycle represented by the current incidence and then to predict the pattern of infection over the next few cycles (5 to 10 years) in the absence of badger removal. Thus four broad categories of outcome might be detected:

(i) alteration/abolition of the periodicity of TB in bovines,
(ii) suppression of the basal underlying rate with retention of the periodicity,
(iii) both of these effects, or
(iv) none of these.

Thus the question as to how the impact of badger removal may be evaluated is not a simple one and depends on whether one is attempting to demonstrate that removal of badgers produces the desired effect overall or whether removal of badgers has had an effect in a particular area. An estimate of the least detectable effect is required, that is to say, that decline in bovine TB after removal of badgers which is significantly different from the current incidence. The effect of badger removal could be compared to an area containing a matched untrapped population of badgers and a matched spectrum of cattle herds. Alternatively the effect of badger removal on the prevailing time-trend within the study area could be examined. These approaches are not mutually exclusive. A further consideration is the number of replications (geographical locations) to be used. Whatever the experimental design, in all cases data on the current incidence together with an estimate of the error (in the statistical sense) associated with this measure are required. Then, given the size of the population of bovines under consideration, it is possible to compute that incidence which is significantly different from the prevailing rate at whatever level of probability is required. The error associated with an estimate of incidence has to take account of methodological limitations of the test system (see Downey 1991).

Effect of trapping in Offaly

The difficulty of establishing a link between removal of badgers and an improvement in the disease status of cattle has already been discussed. Given that the relationship between badger density and TB in badgers has yet to be established and that the concept of a threshold density for maintenance of the disease has some currency, it is most important to consider how successful a removal operation has been and furthermore to consider the likely progression of the recovery of the badger population. From an analysis of the trapping programme in a subsection ($16km^2$) of the clearance area it has been estimated that between 80% and 90% of the badgers were lost to the population over a 2-year period by a combination of trapping and natural mortality. The trapping effort was applied to each social group in two or three sessions in an attempt to reconcile the requirement of adequate trapping at each sett and adequate progress across the site with the constraints of the manpower made available to the project. Nevertheless, given that the catch-curves from Offaly converge to an asymptote, an estimate of the proportion of the population removed is valid. When the survey efficiency is taken into account it may be further estimated that between 70% and 80% of badgers have been removed from the overall site. What is the disposition of those setts which were not trapped? It may be expected that if main setts were missed then the surviving social groups are themselves distributed throughout the clearance area. It follows, then, that these groups are surrounded by vacant territories or by social groups which are severely depleted. The likely fate of these social groups must now be considered. Persistence of these social groups depends on recruitment from the cubs produced therein, by immigration of badgers dispersing from other isolated social groups, or by incorporation of solitary survivors from adjacent social groups. But since the intervening habitat is suitable for badgers and contains vacant setts, which are a valuable resource for badgers (Roper 1993), and since badgers in their social organisation are considered to be essentially contractionist (Kruuk and Macdonald 1985), it could be argued that recruitment by dispersal will not be a major factor contributing to the survival of relatively isolated social groups. Persistence of an isolated group is highly sensitive to random fluctuations in survivorship or mortality. A stochastic model of the population dynamics of social groups of badgers indicates that the mean time to extinction of an isolated social group embedded in surrounding inhospitable habitat is about 10 years. Within 5 years one or other of the sexes will be missing, producing demographic extinction (Lankester *et al.* 1991). If immigration is possible and there are contiguous social groups then the probability of extinction of a particular social group is much reduced. By comparing the initial survey to the intensive study in the 16km^2 sub-area, it seems that of the social groups not detected and trapped about two-thirds are relatively isolated, that is, separated by at least two vacant territories from the nearest undisturbed social group. Since the stochastic model predicts that about half of these will have become extinct within about 4 years, it follows that up to 30% of isolated clans may be removed by these processes. This is certainly an overestimate of the effect of stochastic processes since it does not take account of recruitment by incorporation of solitary survivors of trapped groups and certainly does not adequately take account of long-distance movements by badgers such as have been reported by Sleeman and Mulcahy (1993). Nevertheless it may be concluded that the population has been substantially reduced to less than one badger per km^2 and that it will take at least 5 years for the population to recover. This is close to or below the threshold of one badger per km^2 considered by some to be necessary to sustain TB in bovines (Anderson and Trewhella 1985; McCarthy 1993).

Ecological research is becoming increasingly experimental; hypotheses are constantly being tested by deliberate externally applied disturbances, often with heroic attempts at replication so that standard statistical analyses can be carried out. At the same time much of the time and effort of ecologists is directed towards coping with the analysis of unreplicated events such as the effects of man-made or natural disasters on a local or global scale. The east Offaly study can be and has been criticised for the lack of replication. But it was apparent from the discussion at the seminar that an unpublished consultancy report to the Department of Agriculture and Food, submitted after the operation in east Offaly had commenced, had indicated that an attempt to compare the effect of badger removal with or without specified management initiatives by farmers on TB in cattle would not be feasible on either logistical or cost grounds even before the question of replicates was considered. Thus the analysis of the east Offaly project is of great interest, since it seems unlikely that, unless an unequivocal effect of badger removal is demonstrated, such a widespread control operation will be repeated elsewhere. As described in this seminar by Eves (1993), it seems that the east Offaly project does or could conform to the so-called Before-After-Control-Impact-Pairs design (Stewart-Oaten *et al.* 1992). It has been noted that bovine TB in the control and removal areas has diverged since the project began, although it is too soon to draw any firm conclusions (Eves 1993).

Vaccination of badgers

Attempts to vaccinate badgers against TB, as described in this seminar (Stanford *et al.* 1993; McCarthy 1993), represent a laudable and interesting approach to the problem on two grounds. It may represent a valuable addition to the overall strategy to control bovine TB and if successful will also allow estimation of the contribution of badgers to the bovine problem. How soon might an effect of vaccination of badgers become apparent? Up to 30% of badgers in some areas may be infected, of which about two-thirds have lesions visible at necropsy. The effect of vaccination of these badgers is difficult to predict but is not likely to substantially alter the progress of the disease. Thus the main effect of vaccination would be to reduce or inhibit the entry of badgers uninfected at vaccination to the infected pool, although it might have beneficial effects on some infected badgers (Stanford *et al.* 1993). It seems likely that badgers become infected during the early part of their lifetimes. Age-specific prevalence of TB increases over the first two years of life and pseudovertical transmission from lactating females to their cubs has been reported (Cheeseman, Cresswell *et al.* 1988). Thus younger animals are the main target for vaccination. Given that infected badgers exhibit unusually long survival times and thus are a persistent potential source of cross-infection (Stanford *et al.* 1993), vaccination might not exert an effect until these animals had completed their life-span. The difficulty in identifying such an effect is well illustrated by the data presented in this symposium by McCarthy (1993) where the reactor rate in cattle is oscillatory and the vaccination regime began on the descending phase. It is clearly important for a thorough. analysis of this experiment that facilities be made available to assess the proportion of badgers which have been successfully vaccinated, otherwise it will be impossible fully to evaluate the significance of a negative outcome, i.e. if there is no effect on TB in the area where the vaccine is laid. However, should successful vaccination of badgers prove possible, a realistic estimate of the role of badgers in the maintenance of TB in bovines will follow. An estimate gathered from vaccination studies should be robust since it seems likely that if vaccination reduces TB in badgers then it will have done so in a manner which leaves social organisation and ranging behaviour unaffected. This does not necessarily apply in areas where populations of badgers are reduced by trapping.

Given that the evidence which may quantitatively implicate the badger is still being collected in east Offaly and Cork, one may ask where is the evidence for Ireland that removal of badgers improves TB in cattle? O'Connor and O'Malley (1989) concluded that control of badgers was essential if TB in cattle was to be controlled and they provided data for a number of areas in support of this conclusion. No details of the statistical analyses were given, nor the level of significance attached to the observed decrease in bovine TB after badger removal. The number of badger removals carried out to date must be substantial. This allows criticism of lack of replication to be addressed. Paired control areas can readily be selected from the immense database on TB test results so that such an analysis could be carried out. This matter is currently being addressed by the

Department of Agriculture and Food.

In the meantime it is appropriate to ask how realistic are the estimates of the contribution of the badger to TB in bovines. This was discussed during the seminar and has been specifically addressed by Downey (1991). There are estimated to be about 65,000 animals per year in the Irish cattle herd which are either reactors, infected or both. These include 5000 animals which are infected but unresponsive to the test (false negatives). Such cattle will not be removed by the annual routine test. The potential reactors (60,000) are partitioned between positive infected reactors (50,000) and false positive reactors (10,000). But under recent testing conditions only 75% of the reactors are likely to be identified and removed (Downey 1990). Thus it appears that about 7,500 reactor animals removed annually do not have TB while 12,500 reactors are infected but remain in the herd. Returning to the 50,000 infected cattle arising annually in the herd, the final question to be addressed is what proportion of these can be ascribed to the badger? The answer is not known at present. Nowhere in the written submissions presented in this seminar can such a figure be found. In fact most speakers point out that the role of the badger has not been quantified. Although the contribution of the badger is unknown and could be zero, there was no shortage of scenarios raised during the seminar. In discussion badgers were suggested to be responsible for about 10,000 reactors per annum. This figure, which is based on the proportion of outbreaks considered to be due to the badger and the number of reactors identified in such herds, has also been considered an appropriate estimate of the effect of the badger by Downey (1991). Although this figure has gained some currency, it must be emphasised that it is a hypothesis. This then fuels the debate on the priority to be assigned to addressing the 12,500 reactors estimated to be inadvertently missed by the removal programme or the 2,500 residual reactors (of a total of 10,000) assumed to be caused by badgers which are not removed by the current testing regime. It is clear that far more data on the biology of the badger and the biology of the bovine are required, since if data are produced to partially exonerate one they must at the same time implicate the other or else indicate the presence of hitherto unrecognised factors. It is unlikely that a model of bovine TB in Ireland will be constructed in the near future without additional information on the matrix of interactions within and between cattle and badgers. The case of the possum as a reservoir for bovine TB in New Zealand was raised in discussion as having pedagogical value in our understanding of the role of the badger. A main value of this case is that it indicates that far more of the input parameters to a model of the disease are available for the possum (Morris and Pfeiffer 1990; Barlow 1991 a, b). The details of the analogy should not, however, be forced to excess since the ecological profession of the possum is quite unlike that of the badger (Cowan 1990). Nevertheless deterministic models have succeeded in mimicking the cyclic trend of bovine tuberculosis in badgers, although partitioning the risk of infection between the effects of badgers and cattle has not yet been addressed (Bentil and Murray 1993). For such a complex situation as bovine TB in Ireland, perhaps the main advantages of building simple models are to make explicit the assumptions and the logic of our particular world-view and to identify areas where our knowledge is insufficient.

The application of radio-tracking has produced fundamental insights into the biology of the badger. It is to be expected that the long-awaited computerisation of records of cattle movements will prove equally enlightening. This will allow the impact of cattle movement on TB to be addressed. The Department of Agriculture and Food has been criticised in the past for a reluctance to publish data bearing on the TB problem. This is clearly now being addressed with the establishment of the Tuberculosis Investigation Unit, the publication of information not previously available and the recruitment of epidemiological expertise. It is to be hoped that the current interest in the possibility of control of TB by vaccination of wildlife, requiring as it does a multidisciplinary approach to the problem, may serve as a paradigm for the collegial development of a management plan which realistically and equitably resolves the current conflict between agricultural and conservation interests. Therefore it is to be encouraged.

ACKNOWLEDGEMENTS

The author is extremely grateful to those who participated formally in this seminar for their helpful discussion and clarification of matters arising from their presentations. I am also indebted to the many others who participated in the discussion during and after the symposium. Discussions with Dr P.F.J. Benham, Prof. J.D. Collins, Mr J. Crilly, Dr L. Downey, Ms C. Ferris, Dr R. Hammond and Dr M. Hancox were particularly useful. To the reviewers of the papers (who must remain anonymous) thanks are also due for their constructive criticisms and suggestions.

REFERENCES

ANDERSON, R.M. and TREWHELLA, W. 1985 Population dynamics of the badger (*Meles meles*) and the epidemiology of bovine tuberculosis (*Mycobacterium bovis*) *Philosophical Transactions of the Royal Society of London* **310B**, 327-81.

BARLOW, N.D. 1991a Control of endemic bovine tuberculosis in New Zealand possum populations: results from a simple model. *Journal of Applied Ecology* **28**, 794-800.

BARLOW, N. D. 1991b A spatially aggregated disease/host model for bovine tuberculosis in New Zealand possum populations. *Journal of Applied Ecology* **28**, 777-93.

BARRY, T., POWELL, R. and GANNON, F. 1990 A general method to generate DNA probes for micro-organisms. *Biotechnology* **8**, 233-6.

BENHAM, P.F.J 1993 The interactive behaviour of cattle and badgers with reference to transmission of bovine tuberculosis. In T.J. Hayden (ed.), *The badger,* 189-95. Dublin. Royal Irish Academy.

BENHAM, P.F.J. and BROOM, D.M. 1989 Interactions between cattle and badgers at pasture with reference to bovine tuberculosis transmission. *British Veterinary Journal* **145**, 226-41.

BENHAM, P.F.J. and BROOM, D.M. 1991 Responses of dairy cows to badger urine and faeces in pasture with reference to bovine tuberculosis transmission. *British Veterinary Journal* **147**, 517-32.

BENTIL, D.E. and MURRAY, J.D. 1993 Modelling bovine tuberculosis in badgers. *Journal of Animal Ecology* **62,** 239-50.

BROWN, J., CHEESEMAN, C.L. and HARRIS, S. 1992 Studies on the spread of bovine tuberculosis from badgers to cattle. *Journal of Zoology* **227,** 694-6.
BROWN, J., HARRIS, S. and CHEESEMAN, C.L. 1993 The development of field techniques for studying potential modes of transmission of bovine tuberculosis from badgers to cattle. In T.J. Hayden (ed.), *The badger,* 139-53. Dublin. Royal Irish Academy.
CARR, G. M. and MACDONALD, D. W. 1986 The sociality of solitary foragers: a model based on resource dispersion. *Animal Behaviour* **34**, 1540-9.
CHEESEMAN, C.L. and MALLINSON, P.J. 1979 Radio-tracking in the study of bovine tuberculosis in badgers. In C.I. Amlaner and D.W. Macdonald (eds), *A handbook on biotelemetry and radio tracking,* 649-56. Oxford. Pergamon Press.
CHEESEMAN, C. L., CRESSWELL, W. J., HARRIS, S. and MALLINSON, P.J. 1988 Comparison of dispersal and other movements in two badger (*Meles meles*) populations. *Mammal Review* **18**, 51-9.
CHEESEMAN, C.L., MALLINSON, P.J., RYAN, J. and WILESMITH, J.W. 1993 Recolonisation by badgers in Gloucestershire. In T.J. Hayden (ed.), *The badger,* 78-93. Dublin. Royal Irish Academy.
CHEESEMAN, C.L., WILESMITH, J.W., RYAN, J. and MALLINSON, P.J. 1987 Badger population dynamics in a high density area. *Symposia of the Zoological Society of London* **58**, 279-94.
CHEESEMAN, C. L., WILESMITH, J. W. and STUART, F. A. 1989 Tuberculosis: the disease and its epidemiology in the badger, a review. *Epidemiology and Infection* **103**, 113-25.
COWAN, P. E. 1990 Brushtail possum (*Trichosurum vulpecula* Kerr 1792). In C. H. King (ed.), *The handbook of New Zealand mammals,* 68-98. Oxford University Press.
CRESSWELL, P., HARRIS, S. and JEFFERIES, D.J. 1990 *The history, distribution, status and habitat requirements of the badger in Britain.* Peterborough. Nature Conservancy Council.
DOLAN, L. 1993 Badgers and bovine tuberculosis in Ireland: a review. In T.J. Hayden (ed.), *The badger,* 108-16. Dublin. Royal Irish Academy.
DOWNEY, L. 1990 *Ireland's TB problem: what can and must be achieved?* Dublin. ICOS.
DOWNEY, L. 1991 *Bovine TB programme: what are the realistic expectations?* Dublin. ERAD.
DUNNETT, G.M., JONES, D.M. and MCINERNEY, J.P. 1986 *Badgers and bovine tuberculosis.* London. HMSO.
ELGADI, A. S. and HAYDEN, T.J. 1993 Structural organisation of the badger kidney. In T.J. Hayden (ed.), *The badger,* 123-7. Dublin. Royal Irish Academy.
EVANS, P.G.H., MACDONALD, D.W. and CHEESEMAN, C.L. 1989 Social structure of the Eurasian badger (*Meles meles*): genetic evidence. *Journal of Zoology* **218**, 587-95.
EVES, J. 1993 The East Offaly Badger Research Project: an interim report. In T. J. Hayden (ed.), *The badger,* 166-73. Dublin. Royal Irish Academy.
FAGAN, J. 1993 Tuberculosis in badgers in Ireland: pathology. In T.J. Hayden (ed.), *The badger,* 117-22. Dublin. Royal Irish Academy.

FLANAGAN, P. 1993 Badgers and bovine tuberculosis: epidemiology and case studies. In T.J. Hayden (ed.), *The badger,* 131-8. Dublin. Royal Irish Academy.

FEORE, S. and MONTGOMERY, W.I. 1993 Survey of badger setts in Northern Ireland: progress report. In T.J. Hayden (ed.), *The badger,* 23-5. Dublin. Royal Irish Academy.

GALLAGHER, J. and HORWILL, D.M. 1977 A selective oleic acid albumin agar medium for the cultivation of *Mycobacterium bovis*. *Journal of Hygiene* **79**, 155-60.

GRANGE, J.M., COLLINS, J.D., O'REILLY, L.M., COSTELLO, E. and YATES, M.D. 1990 Identification and characteristics of *Mycobacterium bovis* isolated from cattle, badgers and deer in the Republic of Ireland. *Irish Veterinary Journal* **43**, 33-5.

GRIFFIN, J.M. and HAHESY, T. 1992 Analysis of epidemiology reports on 3957 herd breakdowns in ten DVO regions during 1987-1990. In *Tuberculosis Investigation Unit, University College, Dublin: selected papers 1990-1991*, 21. University College, Dublin.

HARRIS, S. and CRESSWELL, W. J. 1987 Dynamics of a suburban badger (*Meles meles*) population. *Symposium of the Zoological Society of London* **58**, 295-311.

KRUUK, H. 1978 Spatial organisation and territorial behaviour of the European badger (*Meles meles*). *Journal of Zoology London*, **184**, 1-19.

KRUUK, H. 1989 *The social badger: ecology and behaviour of a group-living carnivore* (Meles meles). Oxford University Press.

KRUUK, H. 1993 The organisation of badger populations. In T.J. Hayden (ed.), *The badger,* 4. Dublin. Royal Irish Academy

KRUUK, H. and MACDONALD, D.W. 1985 Group territories of carnivores: empires or enclaves. In R.M. Sibly and R.H. Smith (eds), *Behavioural ecology: ecological consequences of adaptive behaviour*, 521-6. Oxford. Blackwell Scientific Publications.

LANKESTER, K., APELDOORN, R. VAN, MEELIS, E. and VERBOOM, J. 1991 Management perspectives for populations of the Eurasian badger (*Meles meles*) in a fragmented landscape. *Journal of Applied Ecology* **28**, 561-73.

deLISLE, G.W., COLLINS, D.M., LOVEDAY, A.S., YOUNG, W.A.. and JULIAN, A.F. 1990 A report of tuberculosis in cats in New Zealand and the examination of strains of *Mycobacterium bovis* by DNA restriction endonuclease analysis. *New Zealand Veterinary Journal* **38,** 10-13.

LITTLE, T.W.A., NAYLOR, P.F. and WILESMITH, J.W. 1982 Laboratory study of *Mycobacterium bovis* infection in badgers and calves. *Veterinary Record* **111**, 550-7.

LYNCH, J.M., O'CORRY-CROWE, G., CHEESEMAN, C.L., HARRIS, S., DUKE,E.J. and HAYDEN, T.J. 1993 Morphometric and genetic variation among badger populations. In T.J. Hayden (ed.), *The badger,* 94-107. Dublin. Royal Irish Academy.

MCCARTHY, J. 1993 The badger vaccination trial in west Cork: progress report. In T.J. Hayden (ed.), *The badger,* 181-8. Dublin. Royal Irish Academy.

MACDONALD, D.W. 1983 The ecology of carnivore social behaviour. *Nature* **301**, 379-84.

MORRIS, R.S. and PFEIFFER, D. 1990 Report of consultancy on bovine tuberculosis for the Eradication of Bovine Tuberculosis Board (ERAD) (unpublished report) Dublin. ERAD.

MUIRHEAD, R.H., GALLAGHER, J. and BURN, K.J. 1974 Tuberculosis in wild badgers in Gloucestershire: epidemiology. *Veterinary Record* **95**, 552-5.
NOLAN, A. and GOODGER, J. 1993 Diagnosis of tuberculosis in live badgers. In T.J. Hayden (ed.), *The badger,* 128-30. Dublin. Royal Irish Academy.
O'CONNOR, R. and O'MALLEY, E. 1989 *Badgers and bovine tuberculosis in Ireland.* Dublin. ERAD.
O'CORRY-CROWE, G., EVES, J. and HAYDEN, T.J. 1993 Sett distribution, territory size, and population density of badgers (*Meles meles* L.) in east Offaly. In T.J. Hayden (ed.), *The badger,* 35-56. Dublin. Royal Irish Academy.
ROPER, T.J. 1993 Badger setts as a limiting resource. In T.J. Hayden (ed.), *The badger,* 26-34. Dublin. Royal Irish Academy.
SLEEMAN, D.P. and MULCAHY, M.F. 1993 Behaviour of Irish badgers in relation to bovine tuberculosis. In T.J. Hayden (ed.), *The badger,* 154-65. Dublin. Royal Irish Academy.
SMAL, C.M. 1993 The National Badger Survey: preliminary results for the Irish Republic. In T.J. Hayden (ed.), *The badger,* 9-22. Dublin. Royal Irish Academy.
STANFORD, J., STAINSBY, K. and MAHMOOD, K.H. 1993 Could vaccination be the answer to the badger tuberculosis problem? In T.J. Hayden (ed.), *The badger,* 174-80. Dublin. Royal Irish Academy.
STEWART-OATEN, A., BENCE, J.R. and OSENBERG, C.W. 1992 Assessing effects of unreplicated returbations: no simple solutions. *Ecology* **73**, 1396-404.
WARNER, P. 1993 Badgers and the law. In T.J. Hayden (ed.), *The badger,* 5-8. Dublin. Royal Irish Academy.
WHELAN, R. and HAYDEN, T.J. 1993 The reproductive cycle of the female badger (*Meles meles* L.) in east Offaly. In T.J. Hayden (ed.), *The badger,* 64-77. Dublin. Royal Irish Academy.
WOODROFFE, R. and MACDONALD, D.W. 1992 Badger clans: demographic groups in an antisocial species. *Journal of Zoology* **227**, 696-8.